The Floral Artist's Guide
A Reference to Cut Flowers and Foliages

The Floral Artist's Guide
A Reference to Cut Flowers and Foliages

Pat Diehl Scace, AIFD, AAF

DELMAR
™
THOMSON LEARNING

Australia Canada Mexico Singapore Spain United Kingdom United States

DELMAR

THOMSON LEARNING

The Floral Artist's Guide:
A Reference to Cut Flowers and Foliages
by Pat Diehl Scace

Business Unit Director:
Susan L. Simpfenderfer

Executive Editor:
Marlene McHugh Pratt

Acquisitions Editor:
Zina M. Lawrence

Developmental Editor:
Andrea Edwards Myers

Editorial Assistant:
Elizabeth Gallagher

Executive Production Manager:
Wendy A. Troeger

Production Manager:
Carolyn Miller

Project Editor:
Amy E. Tucker

Executive Marketing Manager:
Donna J. Lewis

Channel Manager:
Wendy E. Mapstone

Cover Image:
Courtesy of PhotoDisc.

For permission to use material from this text or product, contact us by
Tel   (800) 730-2214
Fax (800) 730-2215
www.thomsonrights.com

Library of Congress Cataloging-in-Publication Data
Scace, Pat Diehl.
   The floral artist's guide: a reference to cut flowers and foliages / Pat Diehl Scace.
      p. cm.
   ISBN 978-0-7668-1572-8
   1. Cut flowers. 2. Cut foliage. 3. Flower arrangement. I. Title.
SB449.S358 2000
745.92—dc21   00-064506

NOTICE TO THE READER

Publisher does not warrant or guarantee any of the products described herein or perform any independent analysis in connection with any of the product information contained herein. Publisher does not assume, and expressly disclaims, any obligation to obtain and include information other than that provided to it by the manufacturer.

The reader is expressly warned to consider and adopt all safety precautions that might be indicated by the activities herein and to avoid all potential hazards. By following the instructions contained herein, the reader willingly assumes all risks in connection with such instructions.

The Publisher makes no representation or warranties of any kind, including but not limited to, the warranties of fitness for particular purpose or merchantability, nor are any such representations implied with respect to the material set forth herein, and the publisher takes no responsibility with respect to such material. The publisher shall not be liable for any special, consequential, or exemplary damages resulting, in whole or part, from the readers' use of, or reliance upon, this material.

Contents

Preface

Floral design is the expression of oneself and one's emotions through the use of flowers. We use flowers to welcome friends, celebrate special occasions, console loved ones, personalize private spaces, and adorn tables during our most memorable family events.

Have you ever thought about designing flowers? Ever wonder where to begin? One of the most challenging aspects of becoming a floral artist is learning the medium. With hundreds of varied materials commonly used, the task can be overwhelming. This book is meant to aid the reader in learning about those materials both scientifically and in a design capacity. Nearly 500 items of the floral artist's palette are listed here with a pictorial guide and an informational template for each item. The book includes the most typically used, commercially available cut flowers and foliages and some landscape materials that periodically cross over into the cut-flower or greenhouse trade, for use in creating designs or displays. For the beginner, this is a guide to understand the basic identification and naming of plant materials with their general design applications. For the professional designer, it consolidates existing information of those materials into one pictorial and informational resource. The wholesaler will find this book useful for understanding how their customers, i.e., the designer, perceive a product for design applications, further enhancing their sales.

The Floral Artist's Guide, A Reference to Cut Flowers and Foliages is a tool for all artists and flower users to become confident and knowledgeable about their botanical medium, regardless of their design ability.

Acknowledgments

In working on this project I began to reflect on the processes it took for me to get to this point. Spending time in my parents' and grandparents' greenhouses, the late hours during holidays in the flower shop, conversations with my colleagues at the Missouri Botanical Garden, and the vendors and floral "artists" I would see annually at trade functions were all important. Throughout my life I have been blessed to have so many wonderful people encourage and nurture me. This book is a culmination of knowledge gained from every one of those personal and professional relationships. It is my hope that by using this book, others may learn from those experiences.

I want to give special thanks to the following people:

Cliff Willis, Willis Photography, St. Louis, Missouri

Petra Schmidt

John and Bob Baisch, and the staff at Baisch and Skinner, Inc., St. Louis, Missouri

Jerry and Kris Wittenauer, and the staff at Diehl Florist, Inc., Waterloo, Illinois

Dr. Charles and Julie Giedemann

Dr. Marshall R. Crosby

Zina Lawrence, Elizabeth Gallagher, and Andrea Myers at Delmar

Thomas Joseph, Marilyn LeDoux, Jack Jennings, and Mary Beaudin

Most of all to my husband Jeff, children Madison and Dalton, and my entire family and friends.

This book is dedicated to my parents, Ruth B. Diehl Toal and the late LeRoy H. Diehl, for teaching me about the wonderful world of flowers and instilling in me a love for life.

Introduction

Several years ago I left the retail floral business to begin work as Assistant Exhibit Designer for the Missouri Botanical Garden in St. Louis, Missouri. During that time I decided to freelance which provided the opportunity to design for state florist associations, wholesalers, state garden clubs, as well as to train retail employees and teach for the community college. Along the way, I noticed one particular missing thread throughout it all: A comprehensive pictorial identification guide with a template of basic information did not exist to serve all of those varied groups of people; a tool that would provide flower identifications for the amateur and aid the professional in educating the consumer.

The Floral Artist's Guide, A Reference to Cut Flowers and Foliages, is a comprehensive guide, or "palette," which includes commonly used plant material for creating floral works of art. Each entry lists a template of descriptive information:

- Genus name with its pronunciation
- Species, family, and common names
- Related family members stated within the book
- Commercial availability
- Unique characteristics (e.g., fragrance, form, toxicity)
- Design applications of a material highlight their dominant trait (e.g., line, form, filler, or mass) and size (e.g., small, medium, large.)

Additional sections of factual information contain:

- Flower and leaf identification
- Steps to taxonomic identification
- Basic care and handling of perishable products
- Design components
- Index of materials by color
- Name index of Latin and common names

With this information, any user is well prepared to begin the process of selecting materials for successful combinations in design.

How to Use This Book

To better use this book as an identification tool, it is divided into two sections: I. Flowers and Fruits, and II. Foliages and Dried Materials. Flowers are typically the dominant components of a design while foliage and dried materials support those elements. Determine in which section your material will be found, then locate entries alphabetically by their genus (Latin) name. A photograph and a template of information are provided for each item (see template). Use the first six categories for identification information and the last five categories for applications to your designs.

The information at the beginning of this book will help you become familiar with basic botanical information about the materials. With that understanding, the definitions of the design components will help with material applications. Refer to them as necessary when creating your designs.

The name index at the end of the book is helpful if you only know the common name of an item, not the Latin name. Entries are listed alphabetically by common name with a reference to its genus (Latin) name. Most common names in the floral trade are listed.

The color index, also at the end of the book, is helpful if you want to use a specific-colored item but do not know its name. Flowers are loosely grouped into seven categories; foliages and dried materials are loosely grouped into four. Review the list within the desired color and refer to the photo in the book for the material's form. Check the form of your material against the photo and select the appropriate name. Color descriptions vary from grower to wholesaler to consumer, so use the color section as a rough guide. Remember some flowers are available in many colors; for example, *Rosa* has numerous colors and sizes available with frequent additions to the market but photographs of all possibilities are not included here.

If you are unsure of the size of a line, form, filler, or mass flower, then review the entries. Each item states the relative size of a material. Note that sizes of product does vary with peaks in the growing season, and source as determining factors.

2 . . . . *Flowers and Fruit*

Genus Name (Latin): *Aconitum*   **Pronunciation:** (ak-ah-NEE-tum)

Species Name (Latin): *napellus*; others

Common Name: Monkshood, Aconite, *Aconitum*

Family Name (Latin and Common): *Ranunculaceae*, the Crowfoot or Buttercup family

Related Family Members in Book: *Anemone, Clematis, Delphinium, Helleborus,* and *Ranunculas*

Availability: commercial product: April-October

Color: blue

Unique Characteristics: Plant is toxic if ingested. Maintain hand washing during use to prevent possible skin irritation. Hood of sepals is distinctive.

Design Applications: (medium) (line flower) Interesting line element and deep blue color play well against complementary-colored companions in traditional and contemporary designs.

Longevity: days 7-10

Genus Name (Latin): *Aechmea*   **Pronunciation:** (eek-MEE-a)

Species Name (Latin): *fasciata*

Common Name: Urn Plant

Family Name (Latin and Common): *Bromeliaceae*, the Bromelia or Pineapple family

Related Family Members in Book: *Ananas, Guzmania,* and *Tillandsia.*

Availability: Indoor plant has commercial availability year-round.

Color: pink

Unique Characteristics: Foliage and inflorescence combination is distinctive; many other species of *Aechmea* are available, all with varying combinations.

Design Applications: (medium) (form flower) (accent) Excellent choice for adding emphasis to containerized plantings or landscape displays. Provides a nice accent as a cut flower.

Longevity: varies

INFORMATION TEMPLATE DEFINITIONS

Genus Name (Latin): States genus name of material in photograph.

Pronunciation: Lists pronunciation of genus (Latin) name.

Species Name (Latin): If name is bold, species or cultivar is shown in photograph; if not bold it lists commonly used species within the trade; "sp." indicates species is correctly identified but not listed here; "spp." signifies multiple species.

Common Name: Bold first name listed is the most commonly used name by the professional florist. If the first name is a genus name, it is *italicized*. Other common names may also be listed.

Family Name (Latin and Common): Gives the plant family to which the plant belongs.

Related Family Members in Book: Gives other members of the family that are found in this book.

Availability: Gives typical commercial availability from retailers and wholesalers. Items listed as commercial products are available as cut flowers or cut foliage. Landscape plants are those materials available as container plants, for example, annuals, perennials or shrubs. These may also be available from established landscape plants. Indoor plants are those items sold as potted foliage or blooming plants.

Colors: Gives the color of entry if specific species or cultivar is shown. If specific species is not shown, other available colors are listed.

Unique Characteristics: Gives characteristics that are unique to the material, such as fragrance, air-drying capabilities, distinctive form, geotropic or phototropic properties, and toxicity.

Design Applications: Gives relative size and lists primary design applications for the material, for example, line, form, filler, or mass flower; texture; or accent. Possible design options are also listed.

Longevity: Gives typical life expectancy.

How to Use the CD-ROM

The CD-ROM, to accompany *The Floral Artist's Guide: A Reference to Cut Flowers and Foliages,* follows the same format as the book with 507 full-color images divided into two easy-to-browse sections. From the opening screen, you can search the Flowers and Fruits section or choose from numerous foliages and dried materials for your display or arrangement.

Once you've chosen your preferred section, specific options allow you to refine your search by simply clicking from a pull-down menu for the genus, family, species, and even the Common name of the plant. If the name escapes you, or you're matching a particular color scheme, additional features allow you to search for flowers based on color, texture, size, and even longevity! A built-in audio feature allows you to hear the pronunciation of the genus name in Latin. And, information on commercial availability and design applications help you make the proper choices to increase sales and exceed customer expectations.

This handy resource can be used to both test and enhance your product knowledge. Have at-your-fingertips all the specifics you need to know to create stunningly beautiful arrangements on-the-spot for your customers. We provide the resources; the creativity is up to you.

Basic Anatomy of a Flower

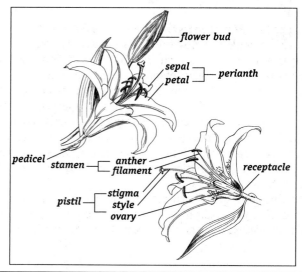

- flower bud
- sepal ┐
- petal ┘ perianth
- pedicel
- stamen ┐ anther
- filament
- pistil ┐ stigma
- style
- ovary
- receptacle

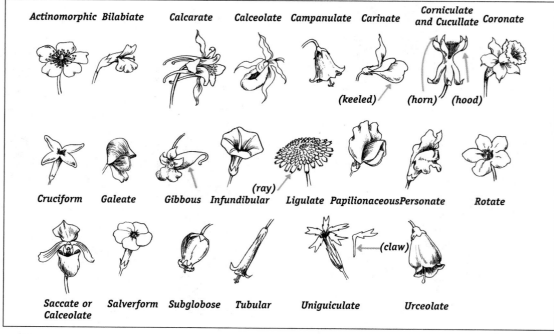

Actinomorphic Bilabiate Calcarate Calceolate Campanulate Carinate Corniculate and Cucullate Coronate

(keeled) (horn) (hood)

Cruciform Galeate Gibbous Infundibular Ligulate Papilionaceous Personate Rotate

(ray)

Saccate or Calceolate Salverform Subglobose Tubular Uniguiculate Urceolate

(claw)

Basic Anatomy of a Leaf

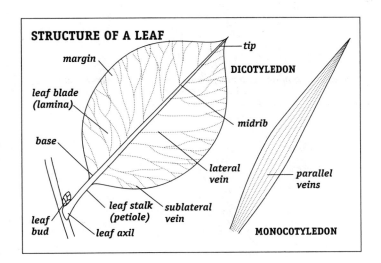

STRUCTURE OF A LEAF

margin

leaf blade
(lamina)

base

leaf
bud

tip

DICOTYLEDON

midrib

lateral
vein

leaf stalk
(petiole)

sublateral
vein

leaf axil

parallel
veins

MONOCOTYLEDON

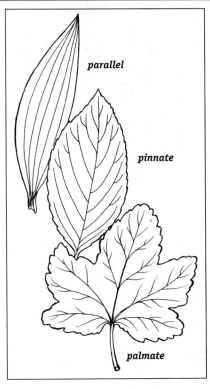

parallel

pinnate

palmate

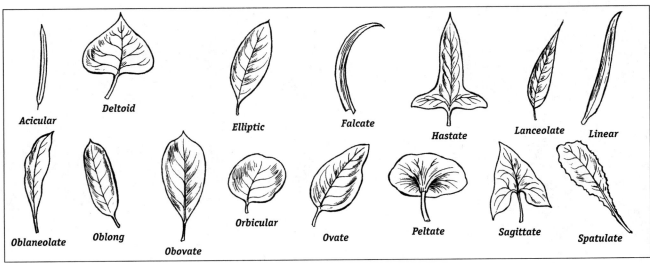

Acicular

Deltoid

Elliptic

Falcate

Hastate

Lanceolate

Linear

Oblaneolate

Oblong

Obovate

Orbicular

Ovate

Peltate

Sagittate

Spatulate

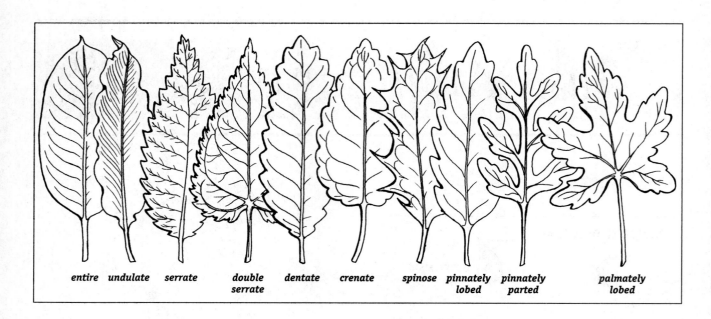

entire undulate serrate double dentate crenate spinose pinnately pinnately palmately
serrate lobed parted lobed

Plant Classification and Nomenclature

Carl Linnaeus, a Swedish naturalist, designed a universal system of plant nomenclature, using a binomial, or two-word, system that classifies plants by their "structural" features. By separating characteristics of a plant and categorizing them into groups, (i.e., classification) similarities and differences can be seen, and the naming process (i.e., nomenclature) begins.

Identification of a plant starts with the genus (genera = plural), or group name. A group with similar characteristics, the species, follows the genus name and is correctly written in italics with the genus capitalized, i.e., *Centaurea macrocephala*. Understanding that both names are based on Latin (sometimes Greek) words, and have specific meaning, will make using the names a more fun and comfortable practice. Again, *Centaurea macrocephala* has a large flowered head: Centaurea is Greek = Kentaur for centaur, which means half man half horse, *macro* meaning "large" and *cephalatus* meaning "bearing heads." Some plants are named after the person who discovered them or for locations of discovery or nativity, for instance, *californicus*. Naturally occuring variants of a species—subspecies, varietas, or forma—are given an additional epithet prefixed by "subsp.," "var.," or "f." Unnatural variants, those forced by man through hybridization, are called 'cultivars' (cultivated variety), and are often named for the person who hybridized the plant. Cultivar name is correctly written in non-italicized type but capitalized and enclosed in single quote marks, i.e., 'Krypton.' Overall family names of a plant consolidate genera by dominant traits; for example, *Euphorbiaceae* includes those plants that secrete a milky sap.

Learning the correct Latin name of flowers is essential to the floral designer because doing so provides a communal language that extends from growers to wholesalers, to retailers and consumers. This better understanding benefits everyone.

The Latin or botanical name of each plant is listed throughout the book. The primary focus here is on the genus name because it is the most commonly used name among the floral trade. Species names are listed only when readily known by the author and are on a separate line to clarify to the beginner that different species within a genus can vary widely in characteristics, even though they share a commonly used name. For example, the differences between *Euphorbia fulgens* and *Euphorbia marginata* are dramatic, yet both are loosely referred to as *Euphorbia* in the trade. The primary source for taxonomic identification in this book was *Hortus Third,* staff of the L. H. Bailey Hortorium, Cornell University. All plants are listed by their current botanical names (except for a few genera, such as *Chrysanthemum,* where an older name has been preserved). Taxonomic names change for many reasons: When research reveals an older name taking precedence, correction of misidentifications, or due to reclassification. These name changes are labeled synonyms, denoted "syn." throughout the book.

Listed below are some species names encountered in this book along with their descriptive meanings.

aconitifolius: having leaves like the aconite; palmately cleft

aethiopicus: Ethiopian; African; south of Libya and Egypt

alatus: winged; having wings or appendages

albidus, albus: white

amabilis: lovable; amiable; lovely

americanus: American

anemoneflorus: with flower like the *Anemone*

anemonefolius: with leaves like the *Anemone*

annuus: annual; living only one year or one plant season

aquaticus, aquatilis: aquatic; living in or under water

aquifolius: holly-leaves; with pointy leaves

argenteus: silvery

armillaris: with bracelet, arm ring, or collar; encircled

aromaticus: aromatic, fragrant

asiaticus: Asian

atropurpureus: dark purple

austriacus: Austrian

balsameus: yielding a fragrant gum or resin

belladonna: beautiful lady

bignonoides: of or like the genus of the trumpet and cross vines

bilobus: having two lobes

bonariensis: of or from Buenos Aires, Argentina

bracteatus: having or bearing bracts

callicarpus: bearing beautiful fruits

campanulatus: campanulate; bell shaped

capsicum: biting to the taste; hot (as peppers)

caudatus: tailed; having a tail-like appendage

coccineus: scarlet

conifer: cone bearing

contortus: twisted

convallis: of the valley

cordatus: heart shaped

cordifolius: having heart-shaped leaves

cristatus: crested; tasseled

deciduous: shedding leaves annually: not evergreen

deliciosus: delicious; of fine flavor; offering great pleasure

divaricatus: spreading at a wide angle; straggling

elegans: elegant

equisetifolius: with leaves like the horsetails, *Equisetum*

ericoides: *Erica*-like; heathlike

flavidus: yellow; yellowish

floribundus: free flowering; abounding in flowers; flowering for a long season

floridus: flowering; full of flowers

fragrans: scented; especially sweet scented

frutescens, fruticans: shrubby, shrublike

gladiatus: like a sword

gloriosus: glorious, superb

gracillimus: very slender

grandiflorus: with large flowers; free flowering

graveolens: heavily scented: strong smelling

herbaceous: not wood forming; of the nature of an herb; low growing—dying back to the ground annually

hirsutus: hairy; covered with coarse hairs

hyacinthus: deep purplish in color

hybridus: mixed; mongrel; crossbred

impatiens: impatient; throwing seed when ripe

incurvus: bent inward; inflexed

integrifolius: with entire or uncut leaves

laevis: smooth; free from hairs or roughness

leonurus: like a lion's tail

leucanthus: with white flowers

linearias: narrow; with nearly parallel sides

luteus: yellow

macranthus: with large thorns or spines

maximus: largest

mollis: soft; flexible; mild

nanus: dwarf

nipponicus: of or from Nippon (Japan)

nobilis: noble; well-known; outstanding

novi-belgii: of New York, United States

occidentalis: western; pertaining to the setting sun

odoratus, odorus: fragrant; scented; sweet smelling

orientalis: eastern; of the dawn

paniculatus: having flowers in a cluster, with each flower borne on a separate stalk

papyrifera: used for producing paper; with paperlike bark

pendulous: hanging down; drooping

podocarpus: bearing fruits on a stalk; literally "fruit-foot"

radicans: with structures that have rooting abilities

rotundifolious: round leaved

rubens, ruber: red; ruddy

salinus: salty; growing in salty places

saxicolus: growing among rocks

scoparius: from the Latin word for "floor-sweeper," resembling a broom

speciosissimus: very showy

sphaerocephalus: round headed

splendens: splendid

stellaris, stellatus: stellate; starry

stolonifera: producing stolons or runners that take root

succulentus: succulent, fleshy, juicy

tenax: tenacious, strong

trilobus: three lobed

unifolious: one leaved

variegatus: variegated

viridis: green

vulgaris, vulgatus: vulgar; common

zebrinus: zebra striped

Tips for Care and Handling of Cut Flowers

The proper care and handling of any cut flower is imperative to its longevity. Proper processing will lengthen the life of materials from 25 to 75 percent. Follow the guidelines below for best results. For the retailer, open the package immediately upon arrival from the supplier and inspect the products for quality or shipping damage. Doing this also allows accumulated ethylene gas released by the flowers to escape. If boxes cannot be processed immediately, review product, then refrigerate the entire package until time allows for correct processing. For the gardener, harvesting flowers from cutting gardens is recommended in the early morning hours and when flowers are in the bud form, breaking or showing color.

PROCESSING GUIDELINES

PREPARE CONTAINERS. Thoroughly clean containers with a commercial cleaner or bleach and brush to remove all existing bacteria along sides, base, and corners.

REMOVE LOWER FOLIAGE. Remove all foliage below the desired water level. Any debris left in water will create bacteria that in turn will clog "xylem" or intake systems. Rinse dirt and debris from base of stems, especially from tulips and other field-grown crops.

RECUT STEMS UNDER WATER. Use a sharp tool to remove the lower 1–3 inches from the base of flower stem. Cut under water to prevent absorption of air bubbles (embolisms) that inhibit water intake.

USE A HYDRATING SOLUTION. Many flowers benefit from dipping their stems into a citric acid–based hydrating solution. Commercial solutions are available and work especially well on "heavy drinkers" such as roses, delphinium, gerberas, and field-grown crops. The solution aids in absorption.

PLACE INTO A PRESERVATIVE SOLUTION. Most commercial preservatives or flower foods are made of dextrose, acids to alter water pH, additives to control bacterial growth, antiethylene substances and other materials to enhance longevity. Follow specified directions for use. If a commercial product is not readily available to the gardener, a few drops of bleach in water will temporarily aid in bacteria reduction. It should not regularly be used as the primary preservative solution. Using warm water produces less air bubbles.

CONDITIONING. Allow the flowers to absorb water at least 30 to 60 minutes to become fully turgid before placing them into refrigeration. Some flowers need overnight conditioning.

REFRIGERATE. Preferred placement of processed flowers is into a cooler with ideal humidity of 80 to 90 percent. Temperature should be 34°F for nontropical flowers; 55°F, or room temperature, for tropicals. Some types of cut foliages benefit from these processing and refrigeration steps. Others are best in a sealed container, needing no additional processing but refrigeration. Always avoid direct sunlight, heating systems, and exposure to freezing temperatures.

CHANGE THE WATER AS NEEDED. Keep containers dirt free, stem tips clean, and water clear. Ensure all flowers receive adequate, clean water. This step is crucial.

ETHYLENE SENSITIVITY

What is ethylene and why should it be a concern? Ethylene is a colorless, odorless gas naturally produced by aging or senescing fruits, flowers, and plants. It is especially prevalent in materials under "stress," for example, materials with physical stress from water loss, diseases such as *botrytis,* or insect infestation. Ethylene is also the by-product of burning natural gas, propane, diesel fuel, and gasoline; it can be found in automobile exhaust or cigarette smoke.

Detrimental effects of this gas on flowers appear as translucent petals, yellowing foliage, or premature death. Correct care and handling procedures will reduce these effects, but is also helpful to pay attention to those flowers which are unusually ethylene sensitive. Limit exposure of such flowers to heavy ethylene producers. Following is a list of flowers known to have some level of ethylene sensitivity:

Achillea	*Freesia*
Aconitum	*Gladiolus*
Agapanthus	*Gypsophila*
Allium	*Iris*
Alstroemeria	*Ixia*
Ammim	*Kniphofia*
Anemone	*Lathyrus*
Anethum	*Lilium*
Antirrhinum	*Matthiola*
Astilbe	*Narcissus*
Bouvardia	*Ornithogalum*
Campanula	*Phlox*
Centaurea	*Physostegia*
Clarkia	*Ranunculas*
Consolida	*Rudbeckia*
Crocosmia	*Saponaria*
Cymbidium	*Scabiosa*
Delphinium	*Solidago*
Dendrobium	*xSolidaster*
Dianthus	*Trachelium*
Digitalis	*Veronica*
Eremurus	

Basics of Floral Design

Design is the aesthetic combination of elements and principles to suit a specific purpose. The materials used are named *elements*. These elements contain components of color, size, line, pattern, form, and texture. In floral design plant material is used as the elements.

Principles are the visual balance, proportion, rhythm, unity, contrast, and harmony of materials being used. When the elements are combined, the secondary principles of accent, repetition, emphasis, scale, tension, and opposition, are produced.

With practice, identifying floral materials by their dominant features will help in applying the elements with the principles to achieve an aesthetically pleasing composition.

This book reinforces the identification of a material by its dominant feature: their uses as line, form, filler, and mass. Doing so strengthens the logical placement within any arrangement. Dominant features as stated in the book are defined as follows:

Consolida

Line Flower—A spike or spikelike inflorescence with an elongated stem. *Consolida, Delphinium, Gladiolus,* or *Antirrhinum* are examples. Foliages or dried materials with linear qualities can also be used to add line.

Stephanotis

Antirrhinum

Freesia

Form Flower—A flower with shape as its most distinctive feature. *Gloriosa, Strelitzia, Stephanotis, Lilium,* or *Iris* are examples. Foliages or dried, such as *Monstera, Philodendron,* or *Physalis,* materials can also add form.

xix

Filler Flower—Any material that is clustered or branched and fills spaces between major components in the design. *Chamelaucium, Gypsophila,* or *Limonium* are examples. Foliages or dried materials can also have filler applications, for example, *Murraya* or *Goniolimon.*

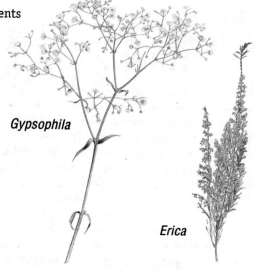

Gypsophila

Erica

Gerbera

Mass Flower—A round flower head on a single stem used for adding visual weight, or mass, to a design, for example, *Dianthus, Rosa, Gerbera,* or *Helianthus.*

The primary principles, secondary principles, and elements—are defined* as follows. Memorize these definitions of principles for applying your elements.

PRIMARY PRINCIPLES

Proportion: The comparative relationship in size, quantity, and degree of emphasis between components within the composition. It is the relationship of one portion to another portion, or the relation of one portion to a whole.

Balance (visual and physical): A state of equilibrium, actual or implied; a feeling of three-dimensional stability. There are two types: symmetrical and asymmetrical. Note: This includes the physical balance of a design when it is carried.

Rhythm: Visual movement through the design usually achieved by means of repetition of line, form, shape, or color.

Contrast: Emphasis by means of difference.

Harmony: Compatibility; the aesthetic quality created by a pleasing interaction of components in a composition.

Unity: Oneness of style, spirit, thought, or purpose; the organization of components into a harmonious whole, resulting in a cohesive relationship of all parts.

*Source: The American Institute of Floral Designers, Book of Floral Terminology.

SECONDARY PRINCIPLES

Accent: Materials or detail added to a design to provide additional interest affecting the totality of the design.

Repetition: The repeating of like elements within a design, for example, line, form, space, color, pattern, texture, or size.

Emphasis: One or several areas within a composition given special attention.

Scale: The relative ratio of size, or the relationship of a composition to the surrounding area or environment.

Tension: Application of materials used to provide an implication or suggestion of opposing forces of energy.

Opposition: Total contrast to induce tension in a design.

Texture: The surface quality of materials as perceived by sight or touch, for example, rough (*Limonium*), smooth (*Gaultheria*), prickly (*Ilex*), velvety (*Sinningia*), and shiny (*Anthurium*).

Focal Zone or **Focal Point:** One or more areas of greatest visual impact or weight; one or more centers of interest to which the eye is drawn to most naturally. Focal point is one emphasized area within the zone of dominance.

ELEMENTS

Color: Visual response of the eye to reflected rays of light; composed of hue, value, and chroma.

Form: The actual shape of an individual component of the composition; the overall three-dimensional configuration, or shape, of a design or composition.

Line: The visual path that directs the eye movement through a composition; the lines may be straight, curved, or a combination; actual or implied.

Pattern: A repeated combination of line, form, color, texture, or space as a single component.

Size: The dimensions of line, form, or space.

Space: The area in, around, and between the design; defined by the three-dimensional totality of the composition; can be negative or positive.

Texture: The surface quality of materials as perceived by sight or touch, for example, rough (*Limonium*), smooth (*Gaultheria*), prickly (*Ilex*), velvety (*Sinningia*), and shiny (*Anthurium*).

Once aware of the basics, follow these helpful hints for creating your designs. Begin by asking yourself a few questions.

Evaluate:
- Evaluate the type of event and location for choosing a size, shape, and style of design for your arrangement. Will your design be for a large room with high ceilings or for a small table in a hospital room? What is the surrounding décor? Is it contemporary or traditional? Will it be viewed from one or all sides?
- Choose a color story, establish a color theme. Will it be complementary or monochromatic? Bold or pastel?
- Choose the container and mechanics appropriately. Will you use a clear glass vase, ceramic container, or a basket? Is it traditional or contemporary? Will it be used for perishable or dried materials? Remember, any material should be adhered securely allowing ample room for water in perishable designs.
- Evaluate each flower for quality. Remove all damaged foliage or unwanted portions.

Create:
- Recut stems on an angle before inserting into desired location. Pay attention to the materials' dominant traits for guidance on placement. Be conscious of the direction and position of stem insertion for your desired style or shape. Should the stem be inserted into a radial or parallel pattern? Will it go in horizontally or vertically?

Use the following line drawings as a reminder of some basic forms and design styles.

A. Fan (Geometric Form)

B. Oval (Geometric Form)

C. Crescent (Geometric Form)

D. Hogarth curve (Geometric Form)

E. Rectangular (Geometric Form)

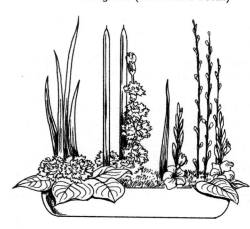

F. Biedermeier (Design Style)

G. Parallel System (Design Style)

H. Vegetative (Design Style)

I. Formal-linear design or Formalinear (Design Style)

J. Abstract (Design Style)

Complete:
- Cover all mechanics. Upon completion, be sure all foam, wire, and tape or means of construction are not visible.

Credits:
All photos were taken by Cliff Willis, Willis Photography, St. Louis, MO. Individual shots by others are listed below.

Amaryllis sp., *Achillea, Allium* sp., *Acanthus, Aesculus, Anemone x hybrida, Astilbe* sp., *Acalypha, Aegopodium, Acer, Adiantum, Ajania, Buddleia, Buxus, Brugmansia, Camellia* spp., *Catalpa, Celosia argentea, Celosia plumosa* 'Apricot Brandy', *Chaenomeles, Cleome, Convallaria, Cotinus, Caladium, Cosmos, Chaemycyparis pisifera* 'Filifera Aurea', *Chasmanthium, Dicentra, Dahlia, Echinacea, Eryngium, Euphorbia pulcherrima, Euonymus fortunei* 'Ivory Jade', *Forsythia, Fagus, Gomphrena, Gladiolus, Hemerocallis, Helleborus, Hydrangea quercifolia, Hydrangea paniculata* 'Tardiva', *Heliotropium, Hosta, Heuchera micrantha* 'Palace Purple', *Ilex opaca, Ilex verticellata, Kolkwitzia, Lavandula, Leucanthemum, Leucadendron, Larix, Liriope spicata, Magnolia grandiflora, Mertensia, Malus* sp., *Mentha, Muscari, Mahonia, Melissa, Miscanthus, Nandina, Nymphaea, Nyssa, Pseudotsuga, Pentas, Paeonia, Perovskia, Physostegia, Pachysandra, Pelargonium, Pennisetum, Phyllostachys, Picea* spp., *Polygonatum, Pyracantha, Rhododendron* sp., *Rudbeckia, Rhus, Rosa floribunda, Sabal, Sedum, Solidago, Salvia* (tricolor), *Senecio cineraria* 'Cirrus', *Stachys, Tsuga, Tradescantia, Verbena bonariensis, Yucca, Zinnia*—Pat Diehl Scace

Vanda—Marilyn LeDoux

Hamamelis, Magnolia stellata—Jack Jennings

Frontmatter Illustrations—Marcia Bales

Botanical Entries

Genus Name (Latin): *Acacia*

Pronunciation: (a-KAY-sha)

Species Name (Latin): sp.

Common Name: *Acacia,* Mimosa, Wattle

Family Name (Latin and Common): *Fabaceae,* the Pea or Pulse family. This family is the second most important source of food for humans and supplies many other economic products.

Related Family Members in Book: *Cercis, Cytisus, Lathyrus,* and *Wisteria*

Availability: commercial product: October-March

Color: yellow

Unique Characteristics: Fuzzy pollen balls are clustered along pendulous stems.

Design Applications: (medium) (filler flower) Displays excellent texture in both flower and foliage, and is useful to add as an accent.

Longevity: days 4-6

Genus Name (Latin): *Acacia*

Pronunciation: (a-KAY-sha)

Species Name (Latin): *retinoides*

Common Name: **Knifeblade Acacia,** Wirilda

Family Name (Latin and Common): *Fabaceae,* the Pea or Pulse family

Related Family Members in Book: *Cercis, Cytisus, Lathyrus,* and *Wisteria*

Availability: commercial product: year-round

Color: yellow

Unique Characteristics: Leaves are narrow and lance-shaped.

Design Applications: (medium) (filler flower) Multiflowered stem is a nice filler; and a good choice for providing textural interest or accent.

Longevity: weeks 1-2

Genus Name (Latin): ***Acanthus***

Pronunciation: (a-KAN-thus)

Species Name (Latin): sp.

Common Name: **Bear's Breeches**, *Acanthus*

Family Name (Latin and Common): *Acanthaceae*, the Acanthus family

Related Family Members in Book: none

Availability: Landscape plant has limited availability as a cut flower.

Color: lavender (shown), white, green, yellow, and pink

Unique Characteristics: Foxglove-like flowers are on 24-inch stems; dark green leaves are sometimes spiny.

Design Applications: (large) (line flower) Unusual flower provides interesting line.

Longevity: days 4-6

Genus Name (Latin): ***Achillea***

Pronunciation: (ah-kil-LEE-ah)

Species Name (Latin): *filipendulina, millefolium;* others and hybrids

Common Name: **Yarrow**, *Achillea*

Family Name (Latin and Common): *Asteraceae*, the Composite or Sunflower family

Related Family Members in Book: *Ageratum, Artemisia, Aster, Centaurea, Chrysanthemum, Cosmos, Craspedia, Dahlia, Echinacea, Echinops-Gerbera, Helianthus, Liatris, Rudbeckia, Senecio, Solidago, x Solidaster,* and *Zinnia*

Availability: commercial product: July-September

Color: yellow (shown), pink, red, and white

Unique Characteristics: Fragrance is sometimes found offensive. It is a good candidate for air-drying, and loses its aroma upon completion.

Design Possibilities: (medium) (mass flower) Works well in mixed bouquets and is useful for terracing. The round flower head is a nice choice for interest in basal focal zones.

Longevity: weeks 1-2 or longer

Genus Name (Latin): **Aconitum**   Pronunciation: (ak-ah-NEE-tum)

Species Name (Latin): *napellus;* others

Common Name: **Monkshood,** Aconite, *Aconitum*

Family Name (Latin and Common): *Ranunculaceae,* the Crowfoot or Buttercup family

Related Family Members in Book: *Anemone, Clematis, Delphinium, Helleborus,* and *Ranunculas*

Availability: commercial product: April-October

Color: blue

Unique Characteristics: Plant is toxic if ingested. Maintain hand washing during use to prevent possible skin irritation. Hood of sepals is distinctive.

Design Applications: (medium) (line flower) Interesting line element and deep blue color play well against complementary-colored companions in traditional and contemporary designs.

Longevity: days 7-10

Genus Name (Latin): **Aechmea**   Pronunciation: (eek-MEE-a)

Species Name (Latin): **fasciata**

Common Name: **Urn Plant**

Family Name (Latin and Common): *Bromeliaceae,* the Bromelia or Pineapple family

Related Family Members in Book: *Ananas, Guzmania,* and *Tillandsia.*

Availability: Indoor plant has commercial availability year-round.

Color: pink

Unique Characteristics: Foliage and inflorescence combination is distinctive; many other species of *Aechmea* are available, all with varying combinations.

Design Applications: (medium) (form flower) (accent) Excellent choice for adding emphasis to containerized plantings or landscape displays. Provides a nice accent as a cut flower.

Longevity: varies

Genus Name (Latin): ***Aesculus***   Pronunciation: (es-KEW-lus)

Species Name (Latin): *parviflora*

Common Name: **Bottlebrush Buckeye,** Dwarf Horsechestnut

Family Name (Latin and Common): *Hippocastanaceae,* the Horsechestnut family

Related Family Members in Book: none

Availability: Landscape plant has flower typically not commercially available.

Color: white

Unique Characteristics: Displays a columnar panicle of florets with stamens exserted.

Design Applications: (large) (line flower) Provides heavy-looking line element.

Longevity: varies

Genus Name (Latin): ***Agapanthus***   Pronunciation: (ag-a-PAN-thus)

Species Name (Latin): *africanus (A. umbellatus), orientalis;* others and hybrids

Common Name: ***Agapanthus,*** Lily-of-the-Nile, African Lily

Family Name (Latin and Common): *Amaryllidaceae,* the Amaryllis family

Related Family Members in Book: *Allium, Amaryllis, Narcissus, Nerine,* and *Triteleia*

Availability: commercial product: March-August.

Color: blue (shown), white

Unique Characteristics: Stem is leafless.

Design Applications: (large) (form flower) Round flower head adds mass to any design style. Leafless flower stem also works well as a line element.

Longevity: days 4-6

Genus Name (Latin): ***Ageratum***

Pronunciation: (aj-ur-AY-tum; a-JER-a-tum)

Species Name (Latin): ***houstonianum* 'Blue Danube';** other ***houstonianum*** cultivars.

Common Name: ***Ageratum***

Family Name (Latin and Common): *Asteraceae,* the Composite or Sunflower family

Related Family Members in Book: *Achillea, Artemisia, Aster, Centaurea, Calendula, Chrysanthemum, Cosmos, Dahlia, Echinops, Echinacea, Liatris, Rudbeckia, Senecio, Solidago, x Solidaster,* and *Zinnia*

Availability: commercial product: June-August

Color: lavender-blue (shown), white, and pink

Unique Characteristics: Flower head has a fuzzy appearance; other cultivars as cut flowers.

Design Applications: (medium) (filler flower) Serves as a nice choice for a unique textural effect, and is a good filler for mixed bouquets.

Longevity: days 4-7

Genus Name (Latin): ***Agrostemma***

Pronunciation: (ag-ro-STEM-a)

Species Name (Latin): sp.

Common Name: ***Agrostemma,* Corn Cockle**

Family Name (Latin and Common): *Caryophyllaceae,* the Pink family

Related Family Members in Book: *Dianthus, Gypsophila,* and *Saponaria*

Availability: commercial product: June-July

Color: purple

Unique Characteristics: Stem has a weeping habit.

Design Applications: (medium) (filler flower) Delicate flower works well in mixed spring bouquets.

Longevity: varies

Genus Name (Latin): ***Ailanthus***   Pronunciation: (ay-LAN-thus)

Species Name (Latin): *altissima*

Common Name: **Tree-of-Heaven,** Varnish Tree

Family Name (Latin and Common): *Simaroubaceae,* the Quassia family

Related Family Members in Book: none

Availability: commercial product: varies

Color: Samara is cream, turning to reddish-orange in maturity.

Unique Characteristics: Samara (winged fruit) is twisted.

Design Applications: (large) (filler) Large stem of unusual fruit adds textural interest.

Longevity: days 10-14

Genus Name (Latin): ***Allium***   Pronunciation: (AL-ee-um)

Species Name (Latin): *giganteum*

Common Name: **Giant Allium,** *Allium*

Family Name (Latin and Common): *Amaryllidaceae,* the Amaryllis family

Related Family Members in Book: *Agapanthus, Amaryllis, Narcissus, Nerine,* and *Triteleia*

Availability: commercial product: April-September

Color: light purple

Unique Characteristics: Stem is leafless with giant orb on top that often smells like onion when cut; odor does not linger after conditioning.

Design Applications: (large) (mass flower) Round flower head adds mass. Leafless stem creates strong vertical lines that work well in parallel or vegetative systems of design.

Longevity: days 5-7

Genus Name (Latin): *Allium*   Pronunciation: (AL-ee-um)

Species Name (Latin): *sphaerocephalon*

Common Name: **Drumstick Allium**

Family Name (Latin and Common): *Amaryllidaceae,* the Amaryllis family

Related Family Members in Book: *Agapanthus, Amaryllis, Narcissus, Nerine,* and *Triteleia*

Availability: commercial product: April-September

Color: light purple

Unique Characteristics: Stem is leafless with small orb on top that often smells like onion when cut; odor does not linger after conditioning.

Design Applications: (small) (form flower) Round flower head adds mass. Leafless flower stem creates strong vertical and horizontal lines that work well in parallel and vegetative systems of design. Nice choice for small contemporary pieces.

Longevity: days 5-7

Genus Name (Latin): *Allium*   Pronunciation: (AL-ee-um)

Species Name (Latin): sp.

Common Name: **Allium**

Family Name (Latin and Common): *Amaryllidaceae,* the Amaryllis family

Related Family Members in Book: *Agapanthus, Amaryllis, Narcissus, Nerine,* and *Triteleia*

Availability: commercial product: Limited

Color: light purple

Unique Characteristics: Stem is leafless; smells like onion when cut; air-dries well.

Design Applications: (large) (mass flower) Round flower head adds mass. Leafless stem creates strong vertical lines that work well in parallel or vegetative systems.

Longevity: days 4-7

Genus Name (Latin): ***Alpinia*** Pronunciation: (al-PIN-ee-ah)

Species Name (Latin): *purpurea, zerumbet;* others

Common Name: **Ginger,** Torch Ginger, Ostrich Plume

Family Name (Latin and Common): *Zingiber-aceae,* the Ginger family

Related Family Members in Book: *Costus, Curcuma, Globba,* and *Zingiber.*

Availability: commercial product: year-round

Color: red, pink (shown)

Unique Characteristics: Flower form is distinctive, with a long-lasting, shiny flower head. Submerging entire stem and flower for 15–20 minutes in lukewarm water will restore appearance if dehydrated.

Design Applications: (large) (form flower) Thick flower stem provides up to 36 inches of length. Stem creates strong vertical or horizontal lines that work best in contemporary or architectural

designs in combination with other tropical flowers.

Longevity: weeks 2-3

Genus Name (Latin): ***Alstroemeria*** Pronunciation: (al-stre-MEAR-ee-ah)

Species Name (Latin): *haemantha, pelegrina;* others and hybrids

Common Name: **Alstroemeria,** Peruvian Lily, Inca Lily, Lily-of-the-Incas

Family Name (Latin and Common): *Alstromeri-aceae,* the Alstroemeria family

Related Family Members in Book: none

Availability: commercial product: year-round

Color: white (shown), yellow, pink, and purple, with most varieties being streaked or freckled with colors

Unique Characteristics: Long-lasting flowers in a variety of colors make this an attractive option for many uses; flower is ethylene sensitive.

Design Applications: (medium) (form flower) Sleek stem with trumpet-shaped flowers is a good specimen for creating the topiary standard.

Longevity: days 12-15

Genus Name (Latin): *Amaranthus*

Pronunciation: (am-a-RAN-thus)

Species Name (Latin): *caudatus*

Common Name: ***Amaranthus,*** Love-Lies-Bleeding, Tassel Flower

Family Name (Latin and Common): *Amaranthaceae,* the Amaranth family

Related Family Members in Book: *Celosia, Gomphrena*

Availability: commercial product: July-October

Color: red

Unique Characteristics: Inflorescence will air-dry well, losing only slight color and many small seeds.

Design Applications: (medium) (line flower) (texture) Displays a coarse appearance in pendent panicles, and serves as a useful material for adding vertical lines of textural interest; works well in combination with autumnal-colored flowers.

Longevity: days 6-10 or longer

Genus Name (Latin): *Amaranthus*

Pronunciation: (am-a-RAN-thus)

Species Name (Latin): *caudatus* 'Viridis'

Common Name: ***Amaranthus,*** Love-Lies-Bleeding, Tassel Flower

Family Name (Latin and Common): *Amaranthaceae,* the Amaranth family

Related Family Members in Book: *Celosia, Gomphrena*

Availability: commercial product: July-October

Color: green

Unique Characteristics: Inflorescence will air-dry well, losing only some color and many seeds.

Design Applications: (medium) (line flower) (texture) Displays a coarse appearance in pendent panicles, and works well along container edges, arching armatures, or any position from which the flowers can freely be suspended. Sometimes can also be a useful filler.

Longevity: days 6-10 or longer

Genus Name (Latin): **Amaryllis**   Pronunciation: (am-a-RIL-is)

Species Name (Latin): sp.

Common Name: **Belladonna Lily,** Naked Lady Lily, Cape Belladonna

Family Name (Latin and Common): *Amaryllidaceae,* the Amaryllis family

Related Family Members in Book: *Agapanthus, Allium, Narcissus, Nerine,* and *Triteleia*

Availability: Commercial product has limited availability: May-November

Color: pinks, red, and white

Unique Characteristics: Stem is hollow with trumpet-shaped flowers, 2–3 inches across.

Design Applications: (medium) (form flower) Trumpet-shaped flowers on hollow stem provide strong vertical line. Stem is well suited for parallel and vegetative systems of design.

Longevity: days 7-14, with individual florets continuing to open

Genus Name (Latin): **Ammi**   Pronunciation: (AM-mee)

Species Name (Latin): *majus*

Common Name: **Queen Anne's Lace,** Bishop's Weed

Family Name (Latin and Common): *Apiaceae,* the Parsley or Carrot family

Related Family Members in Book: *Anethum, Bupleurum, Eryngium,* and *Trachymene*

Availability: commercial product: year-round

Color: white

Unique Characteristics: Form is similar to dill and to species of wild carrot.

Design Applications: (large) (filler flower) Airy flower heads work well in mass arrangements.

Longevity: days 3-5

Genus Name (Latin): ***Ananas***

Pronunciation: (a-NAN-as; a-NAY-nas)

Species Name (Latin): sp.

Common Name: **Dwarf Ornamental Pineapple**

Family Name (Latin and Common): *Bromeli-aceae,* the Bromelia or Pineapple family

Related Family Members in Book: *Aechmea, Guzmania,* and *Tillandsia*

Availability: commercial product: year-round

Color: white and pink variegated

Unique Characteristics: Form resembles edible pineapple.

Design Applications: (small) (form flower) Serves as an excellent choice for Hawaiian theme designs. Size allows for varying scaled arrangements; works best in combination with tropical flowers.

Longevity: weeks 1-3 long lasting

Genus Name (Latin): ***Ananas***

Pronunciation: (a-NAN-as; a-NAY-nas)

Species Name (Latin): sp.

Common Name: **Ornamental Pineapple**

Family Name (Latin and Common): *Bromeli-aceae,* the Bromelia or Pineapple family

Related Family Members in Book: *Aechmea, Guzmania,* and *Tillandsia*

Availability: commercial product: year-round

Color: white and pink variegated

Unique Characteristics: Form resembles edible pineapple.

Design Applications: (medium) (form flower) Serves as an excellent choice for Hawaiian theme designs. Size allows for varying scaled arrangements; works best in combination with tropical flowers.

Longevity: weeks 1-3 long lasting

Genus Name (Latin): ***Anemone***   Pronunciation: (a-NEM-oh-nee)

Species Name (Latin): ***coronaria*** cultivar; others and hybrids

Common Name: ***Anemone,*** Poppy Anemone, Windflower, Lily-of-the-Field, (florist's anemone)

Family Name (Latin and Common): *Ranunculaceae,* the Crowfoot or Buttercup family

Related Family Members in Book: *Clematis, Delphinium, Helleborus, Nigella,* and *Ranunculas*

Availability: commercial product: October-May

Color: purple, dark blue, pink, red (shown), and white

Unique Characteristics: Some *Anemones* are phototropic, or bend toward light; prefers cooler temperatures.

Design Applications: (small) (mass flower) Bold colored petals surrounding a solid black center make a dramatic statement. Nice specimen to use in massed bunches.

Longevity: days 3-7

Genus Name (Latin): ***Anemone***   Pronunciation: (a-NEM-oh-nee)

Species Name (Latin): ***x hybrida,*** others and cultivars

Common Name: **Japanese Anemone**

Family Name (Latin and Common): *Ranunculaceae,* the Crowfoot or Buttercup family

Related Family Members in Book: *Clematis, Delphinium, Helleborus, Nigella,* and *Ranunculas*

Availability: Landscape plant is limited as a cut flower.

Color: white (shown), rose, and crimson

Unique Characteristics: Plant is phototropic, or bends toward light. Longevity is best when the flower is cut from the plant during early-morning hours.

Design Applications: (medium) (filler flower) Multiple buds on stem work well as a filler flower; nice choice for mixed garden bouquets.

Longevity: days 5-7 or longer

Genus Name (Latin): **Anethum**   Pronunciation: (a-NAY-thum)

Species Name (Latin): *graveolens*

Common Name: **Dill**

Family Name (Latin and Common): *Apiaceae,* the Parsley or Carrot family

Related Family Members in Book: *Ammi, Bupleurum, Eryngium,* and *Trachymene*

Availability: commercial product: year-round

Color: green

Unique Characteristics: Flower form is similar in appearance to Queen Ann's Lace, on stems often 30-36 inches long. Used in cooking because the plant has a strong scent.

Design Applications: (large) (filler flower) Yellow-green color, combined with its significance as an herb, works well with designs created using fresh vegetables; also is a nice choice for filler flower in mass arrangements; especially useful in parallel designs.

Longevity: days 7-10

Genus Name (Latin): **Anigozanthus**   Pronunciation: (a-nee-go-ZAN-thus)

Species Name (Latin): *flavidus, pulcherrimus, rufus;* others and cultivars

Common Name: **Kangaroo Paw**

Family Name (Latin and Common): *Haemodoraceae,* the Bloodwort family

Related Family Members in Book: none

Availability: commercial product: year-round

Color: red, yellow, and green

Unique Characteristics: Flower air-dries easily and is hirsute.

Design Applications: (small) (form flower) Unique appearance works well with grasses and other exotics.

Longevity: days 10-14

Genus Name (Latin): ***Anthurium***   Pronunciation: (an-THEW-ree-um)

Species Name (Latin): ***andreanum*** hybrid; many others

Common Name: ***Anthurium,*** Flamingo Lily

Family Name (Latin and Common): *Araceae,* the Arum family

Related Family Members in Book: *Caladium, Monstera, Philodendron,* and *Zantedeschia*

Availability: commercial product: year-round

Color: red (shown), pink, purple, and white

Unique Characteristics: Form is distinctive; color is in the spathe; has a waxy-looking surface; and is best if stored above 45°F and misted often with water.

Design Applications: (medium) (form flower) Distinctive sleek, heart-shaped form provides instant drama; works best in combination with other tropical or exotic flowers.

Longevity: weeks 1-2

Genus Name (Latin): ***Anthurium***   Pronunciation: (an-THEW-ree-um)

Species Name (Latin): ***andreanum*** hybrid; many others

Common Name: ***Anthurium,*** Flamingo Lily

Family Name (Latin and Common): *Araceae,* the Arum family

Related Family Members in Book: *Caladium, Monstera, Philodendron,* and *Zantedeschia*

Availability: commercial product: year-round

Color: white with pink venation (shown), red, purple, and white

Unique Characteristics: Form is distinctive; color is in the spathe; has a waxy-looking surface; and is best if stored above 45°F and misted often with water.

Design Applications: (medium) (form flower) Distinctive sleek, heart-shaped form provides instant drama; works best in combination with other tropical or exotic flowers.

Longevity: weeks 1-2

Genus Name (Latin): ***Anthurium***   Pronunciation: (an-THEW-ree-um)

Species Name (Latin): ***andreanum*** hybrid; many others

Common Name: ***Anthurium,*** Flamingo Lily

Family Name (Latin and Common): *Araceae,* the Arum family

Related Family Members in Book: *Caladium, Monstera, Philodendron,* and *Zantedeschia*

Availability: commercial product: year-round

Color: red mottled white (shown), red, purple, pink, and white

Unique Characteristics: Form is distinctive; color is in the spathe; has a waxy-looking surface; and is best if stored above 45°F and misted often with water.

Design Applications: (medium) (form flower) Distinctive sleek, heart-shaped flower provides instant drama; works best with other tropical and exotic flowers.

Longevity: weeks 1-2

Genus Name (Latin): ***Anthurium***   Pronunciation: (an-THEW-ree-um)

Species Name (Latin): ***andreanum*** hybrid; many others

Common Name: ***Anthurium,*** Flamingo Lily

Family Name (Latin and Common): *Araceae,* the Arum family

Related Family Members in Book: *Caladium, Monstera, Philodendron,* and *Zantedeschia*

Availability: commercial product: year-round

Color: red (shown), purple, white, red and white mottled

Unique Characteristics: Form is distinctive and miniature in size at 3 inches across; color is in the spathe; has a waxy-looking surface; and is best if stored above 45°F and misted often with water.

Design Applications: (medium) (form flower) Distinctive sleek, heart-shaped flower provides instant drama; works best with other tropical and exotic flowers.

Longevity: weeks 1-2

Genus Name (Latin): *Anthurium*

Pronunciation: (an-THEW-ree-um)

Species Name (Latin): 'Krypton'

Common Name: *Anthurium,* Flamingo Lily

Family Name (Latin and Common): *Araceae,* the Arum family

Related Family Members in Book: *Caladium, Monstera, Philodendron,* and *Zantedeschia*

Availability: greenhouse plant: year-round

Color: pink

Unique Characteristics: Flower form is distinctive and miniature in size at 1-2 inches across; works well as an accent plant and for cut flowers and foliage use.

Design Applications: (small) (form flower) Smaller size than traditional *Anthurium,* this flower is an excellent choice for unusual corsages, boutonnieres, and bridal bouquets as well as container designs.

Longevity: days 10-14 as cut flower

Genus Name (Latin): *Anthurium*

Pronunciation: (an-THEW-ree-um)

Species Name (Latin): *andreanum* hybrid; many others

Common Name: *Anthurium,* Flamingo Lily

Family Name (Latin and Common): *Araceae,* the Arum family

Related Family Members in Book: *Caladium, Monstera, Philodendron,* and *Zantedeschia*

Availability: commercial product: year-round

Color: purple

Unique Characteristics: form is distinctive, as the heart-shaped leaf (spathe) of this hybrid is reflexed; best if stored above 45°F and misted often with water.

Design Applications: (small) (form flower) Flower is best used in combination with other exotics or in mass by itself; it is often used in contemporary or geometric styles.

Longevity: weeks 1-2

Genus Name (Latin): ***Antirrhinum***   Pronunciation: (an-tee-RYE-num)

Species Name (Latin): ***majus*** cultivar; many others

Common Name: **Snapdragon**

Family Name (Latin and Common): *Scrophulari-aceae,* the Figwort family

Related Family Members in Book: *Digitalis, Leucophyllum, Penstemon,* and *Veronica*

Availability: commercial product: year-round

Color: pink (shown), white, yellow, purple, and orange

Unique Characteristics: Flower stem is geotropic in that stems bend away from gravity. Store in vertical position to avoid curvature. Individual florets "snap" when gently squeezed. Most varieties have sweet fragrance. Due to ethylene sensitivity, avoid prolonged packaging in airtight containers.

Design Applications: (medium) (line flower) Serves as an excellent choice for any style of design.

Longevity: days 10-14

Genus Name (Latin): ***Aquilegia***   Pronunciation: (ak-wi-LEE-ji-a)

Species Name (Latin): sp.

Common Name: **Columbine**

Family Name (Latin and Common): *Ranuncu-laceae,* the Crowfoot or Buttercup family

Related Family Members in Book: *Aconitum, Anemone, Clematis, Delphinium, Helleborus, Nigella,* and *Ranunculas*

Availability: Landscape plant has varied commercial availability as a cut flower, April-September.

Color: red-yellow bicolor (shown), blue, pink, purple, yellow, and white (all predominantly bicolor)

Unique Characteristics: Flower form often contains spurred petals.

Design Applications: (small) (form flower) Distinctive hooks or spurs on petals and unique flower shape are excellent additions to vegetative or garden-style designs.

Longevity: varies

Genus Name (Latin): ***Arachnis***

Pronunciation: (a-RAK-nis)

Species Name (Latin): hybrids

Common Name: ***Arachnis,*** Spider Orchid, Scorpion Orchid

Family Name (Latin and Common): *Orchidaceae,* the Orchid family

Related Family Members in Book: *Cattleya, Dendrobium, Oncidium, Paphiopedilum, Cymbidium, Phalaenopsis,* and *Vanda*

Availability: commercial product: year-round

Color: red

Unique Characteristics: Flower form has the appearance of a spider; and sometimes is fragrant.

Design Applications: (small) (form flower) Serves as an excellent choice for adding drama, and is most effective if allowed enough space to display its unique character.

Longevity: days 7-10

Genus Name (Latin): ***Asclepias***

Pronunciation: (as-KLEE-pee-us)

Species Name (Latin): ***tuberosa***

Common Name: ***Asclepias,*** Indian Paintbrush, Butterfly Weed

Family Name (Latin and Common): *Asclepi-adaceae,* the Milkweed family

Related Family Members in Book: *Stephanotis, Tweedia*

Availability: commercial product: year-round, with peak supplies from June- September

Color: orange

Unique Characteristics: Stem exudes milky sap. Dipping stems into hot water upon cutting deters continual sap flow.

Design Applications: (medium) (filler flower) Clusters of umbellate florets continue to open, creates visual weight. Provides an excellent color mass when placed in focal zone.

Longevity: days 3-7

Genus Name (Latin): *Asclepias*   Pronunciation: (as-KLEE-pee-us)

Species Name (Latin): spp.

Common Name: **Asclepias pods**

Family Name (Latin and Common): *Asclepiadaceae,* the Milkweed family

Related Family Members in Book: *Stephanotis, Tweedia*

Availability: commercial product: varies, late summer-early fall

Color: green (light)

Unique Characteristics: Hairy, densely covered, multiple pods are attached to the stem.

Design Applications: (medium) (form flower) Use in designs where unusual form is highly visible.

Longevity: days 5-7

Genus Name (Latin): *Aster*   Pronunciation: (AS-ter)

Species Name (Latin): *ericoides* '**Monte Cassino'**

Common Name: **Monte Cassino Aster**

Family Name (Latin and Common): *Asteraceae,* the Composite or Sunflower family

Related Family Members in Book: *Achillea, Ageratum, Calendula, Centaurea, Chrysanthemum, Cosmos, Craspedia, Dahlia, Helianthus, Liatris, Solidago, Tagetes,* and *Zinnia*

Availability: commercial product: June-September

Color: white

Unique Characteristics: Flower has prominent yellow center.

Design Applications: (small) (filler flower) Daisylike flowers lend themselves to garden-style designs; flowers are useful as filler.

Longevity: days 7-10

Genus Name (Latin): ***Aster*** Pronunciation: (AS-ter)

Species Name (Latin): ***novi-belgii***

Common Name: **Novi-Belgii Aster**

Family Name (Latin and Common): *Asteraceae,* the Composite or Sunflower family

Related Family Members in Book: *Achillea, Ageratum, Calendula, Centaurea, Chrysanthemum, Cosmos, Craspedia, Dahlia, Helianthus, Liatris, Solidago, Tagetes,* and *Zinnia*

Availability: commercial product: July-December

Color: violet

Unique Characteristics: Flower has prominent yellow center.

Design Applications: (small) (filler flower) Daisylike flowers lend themselves to garden-style designs. This type works well in combination with shades of pink and deep purple, as a useful filler.

Longevity: days 7-10

Genus Name (Latin): ***Astilbe*** Pronunciation: (a-STIL-bee)

Species Name (Latin): sp.

Common Name: **Astilbe,** False Spiraea, Goat's Beard, Meadowsweet

Family Name (Latin and Common): *Saxifragaceae,* the Saxifrage family

Related Family Members in Book: *Heuchera*

Availability: commercial product: March-December

Color: pinks, red, and white

Unique Characteristics: Flower form is pyramidal, with a feathery texture; florist's astilbes are likely to be hybrids.

Design Applications: (medium) (filler flower) Unique filler will also add mass; works well in combination with other spring flowers.

Longevity: days 4-7

Genus Name (Latin): ***Banksia***   Pronunciation: (BANK-see-uh)

Species Name (Latin): ***coccinea***

Common Name: **Scarlet Banksia**

Family Name (Latin and Common): *Proteaceae,* the Protea family

Related Family Members in Book: *Leucadendron, Leucospermum, Protea, Serruria,* and *Telopea*

Availability: commercial product: December-April

Color: orange with gray undertones

Unique Characteristics: Inflorescence (flower) is actually several hundred tiny flowers; the combination of colors and texture within the specimen is unique to the species; air-dries well after removing from water source.

Design Applications: (medium) (form flower) Focal zones are the ideal location for all *Banksia.* Serves as an excellent choice for adding accent.

Longevity: weeks 2-3 long lasting

Genus Name (Latin): ***Banksia***   Pronunciation: (BANK-see-uh)

Species Name (Latin): ***hookeriana***

Common Name: **Hookeriana Banksia**

Family Name (Latin and Common): *Proteaceae,* the Protea family

Related Family Members in Book: *Leucadendron, Leucospermum, Protea, Serruria,* and *Telopea*

Availability: commercial product: February-April

Color: orange

Unique Characteristics: Flower air-dries well; form is distinctive with base of flower head containing densely matted, short, woolly hairs; more rounded form than other species in book.

Design Applications: (large) (form flower) This dramatic flower works well for focal zones. Foliage can also be used by itself for adding interest.

Longevity: weeks 2-3 long lasting

Genus Name (Latin): ***Banksia***   Pronunciation: (BANK-see-uh)

Species Name (Latin): ***menziesii***

Common Name: **Raspberry Frost Banksia**

Family Name (Latin and Common): *Proteaceae*, the Protea family

Related Family Members in Book: *Leucadendron, Leucospermum, Protea, Serruria,* and *Telopea*

Availability: commercial product: August-April

Color: green-yellow with pale pink center

Unique Characteristics: Flower air-dries well; form is distinctive, with base of flower head containing densely matted, short, woolly hairs.

Design Applications: (large) (form flower) Flower lends itself to focal zone placement. Unique foliage creates striking texture.

Longevity: weeks 2-3 long lasting

Genus Name (Latin): ***Banksia***   Pronunciation: (BANK-see-uh)

Species Name (Latin): ***prionotes***

Common Name: **Acorn Banksia**

Family Name (Latin and Common): *Proteaceae*, the Protea family

Related Family Members in Book: *Leucadendron, Leucospermum, Protea, Serruria,* and *Telopea*

Availability: commercial product: May-October

Color: orange, with top portion much lighter

Unique Characteristics: Flower air-dries well; flower form is woolly and can be more elongated than other *Banksia* listed in this book.

Design Applications: (large) (form flower) Dramatic flower adds accent like other *Banksia*. Foliage provides excellent textural interest.

Longevity: weeks 2-3 long lasting

Genus Name (Latin): ***Banksia*** Pronunciation: (BANK-see-uh)

Species Name (Latin): sp.

Common Name: **Banksia**

Family Name (Latin and Common): *Proteaceae,* the Protea family

Related Family Members in Book: *Leucadendron, Leucospermum, Protea, Serruria,* and *Telopea*

Availability: commercial product: **year-round**

Color: **yellowish-orange**

Unique Characteristics: Flower air-dries well.

Design Applications: (large) (form flower) (accent) Like most *Banksia,* focal-point placements are most dramatic. Excellent choice for adding accent.

Longevity: weeks 2-3 long lasting

Genus Name (Latin): ***Banksia*** Pronunciation: (BANK-see-uh)

Species Name (Latin): ***speciosa***

Common Name: **Banksia,** Rick Rack

Family Name (Latin and Common): *Proteaceae,* the Protea family

Related Family Members in Book: *Leucadendron, Leucospermum, Protea, Serruria,* and *Telopea*

Availability: commercial product: **March-August**

Color: **green (light)**

Unique Characteristics: Color

Design Applications: (large) (form flower) Serves as an excellent choice for accent, working uniquely well with ornamental grasses.

Longevity: weeks 2-3 long lasting

Genus Name (Latin): ***Begonia***   Pronunciation: (bee-GO-nee-uh)

Species Name (Latin): spp.

Common Name: **Tuberous Begonia**

Family Name (Latin and Common): *Begoniaceae,* the Begonia family

Related Family Members in Book: none

Availability: Greenhouse plant has varied commercial availability.

Color: pink, white, yellow, orange, red, and apricot

Unique Characteristics: Flowers vary among cultivars with interesting combinations of foliage.

Design Applications: (medium) (form) Flowering plant is useful for adding accent to European gardens and containerized plantings.

Longevity: varies

Genus Name (Latin): ***Begonia***   Pronunciation: (bee-GO-nee-uh)

Species Name (Latin): **x semperflorens-cultorum** hybrid

Common Name: **Begonia** (double)

Family Name (Latin and Common): *Begoniaceae,* the Begonia family

Related Family Members in Book: none

Availability: Landscape plant has varied availability during spring and summer.

Color: pinks, reds, and white

Unique Characteristics: Foliage looks waxy; has a double petal flower. Typically used as a bedding plant.

Design Applications: (small) (form) Plant is a good choice for adding textural contrast in European gardens or other containerized plantings.

Longevity: varies

Genus Name (Latin): ***Begonia***   Pronunciation: (bee-GO-nee-uh)

Species Name (Latin): **x semperflorens-cultorum** hybrid

Common Name: **Begonia** (single)

Family Name (Latin and Common): *Begoniaceae,* the Begonia family

Related Family Members in Book: none

Availability: Landscape plant has varied availability during spring and summer.

Color: reds, pinks, and white

Unique Characteristics: Foliage looks waxy; flower has a yellow center; typically used as a bedding plant.

Design Applications: (small) (form) Plant is a good choice for adding textural contrast in European gardens or other containerized plantings.

Longevity: varies

Genus Name (Latin): ***Boronia***   Pronunciation: (bor-OH-nee-uh)

Species Name (Latin): ***heterophylla***

Common Name: ***Boronia***

Family Name (Latin and Common): *Rutaceae,* the Rue family.

Related Family Members in Book: *Diosma*

Availability: commercial product: January-May

Color: hot pink

Unique Characteristics: Color is bold, and so works well in combination with other pinks or contrasting colors. Flower has a unique citruslike fragrance, and is ethylene sensitive.

Design Applications: (small) (filler flower) Unique texture and bold color of this filler work well in traditional or contemporary designs.

Longevity: days 5-7

Genus Name (Latin): ***Bougainvillea***   Pronunciation: (boo-gin-vil-EE-ah)

Species Name (Latin): spp.

Common Name: ***Bougainvillea***

Family Name (Latin and Common): *Nyctaginaceae,* the Four-o'clock family

Related Family Members in Book: none

Availability: Landscape plant has limited availability during spring and summer as a cut flower.

Color: magenta, light purple, yellow, and white

Unique Characteristics: Colorful bracts are paperlike.

Design Applications: (small) (form flower) Known for its bright colors, the trailing habit works well in tropical or exotic designs.

Longevity: days 3-4 as a cut flower

Genus Name (Latin): ***Bouvardia***   Pronunciation: (boo-VAR-dee-uh)

Species Name (Latin): spp.

Common Name: ***Bouvardia*** (double)

Family Name (Latin and Common): *Rubiaceae,* the Madder family

Related Family Members in Book: *Gardenia, Pentas*

Availability: commercial product: year-round

Color: pink, white

Unique Characteristics: Flower head has double-petaled, star-shaped florets; longevity is best if used in clear water, not floral foam; this flower is ethylene sensitive.

Design Applications: (medium) (mass flower) Cluster of florets is useful for adding mass; works uniquely well in hand-tied bouquets.

Longevity: days 10-14

Genus Name (Latin): *Bouvardia*   Pronunciation: (boo-VAR-dee-uh)

Species Name (Latin): spp.

Common Name: **Bouvardia** (single)

Family Name (Latin and Common): *Rubiaceae,* the Madder family

Related Family Members in Book: *Gardenia, Pentas*

Availability: commercial product: year-round

Color: pink, white

Unique Characteristics: Flower head has delicate star-shaped florets that create almost a flat-top appearance; longevity is best if used in clear water, not floral foam; this flower is ethylene sensitive.

Design Applications: (medium) (mass flower) Cluster of florets is useful for adding mass; works uniquely well in hand-tied bouquets.

Longevity: days 10-14

Genus Name (Latin): *Bracteantha* syn. *Helichrysum bracteatum*   Pronunciation: (brak-tee-ANTH-a)

Species Name (Latin): **bracteata** cultivar

Common Name: **Strawflower**

Family Name (Latin and Common): *Asteraceae,* the Composite or Sunflower family

Related Family Members in Book: *Achillea, Ageratum, Artemisia, Aster, Centaurea, Chrysanthemum, Cosmos, Craspedia, Dahlia, Echinops, Echinacea, Gerbera, Helianthus, Liatris, Rudbeckia, Senecio, Solidago, x Solidaster,* and *Zinnia*

Availability: commercial product: varies October-November

Color: yellow, orange, white, red, and pink

Unique Characteristics: Flower air-dries well, and has strawlike appearance. Stem stays strong on fresh flower but will need wire support when it becomes dry.

Design Applications: (small) (mass flower) Serves as an excellent choice for adding texture to autumnal bouquets; and is useful as a mass flower in mixed arrangements.

Longevity: days 4-7

Genus Name (Latin): ***Brassica***

Pronunciation: (BRASS-i-kuh)

Species Name (Latin): *oleracea* cultivar

Common Name: **Ornamental Kale,** Wild Cabbage

Family Name (Latin and Common): *Brassicaceae,* the Mustard family

Related Family Members in Book: *Lunaria, Matthiola*

Availability: Landscape plant has varied commercial availability.

Color: white, pinks, and red-purple

Unique Characteristics: Foliage has rosette form, with some containing ruffled or feathery margins.

Design Applications: (large) (mass flower) Rosettes work as "flowers" in temporary landscape displays. Individual leaves add both mass and wonderful texture to any cut-flower design.

Longevity: varies as a cut foliage; weeks when used as a bedding plant, if temperatures remain cool

Genus Name (Latin): ***Brugmansia***

Pronunciation: (brug-MAN-zee-a)

Species Name (Latin): sp.

Common Name: **Angel's Trumpets**

Family Name (Latin and Common): *Solanaceae,* the Nightshade family

Related Family Members in Book: *Capsicum, Physalis*

Availability: varies commercially as a greenhouse or bedding plant

Color: white (shown), yellow, orange, and pink

Unique Characteristics: Plant is fragrant at night; has pendent, tubular, trumpet-shaped flowers.

Design Applications: (large) (form) Serves as an excellent choice for adding accent to containerized gardens.

Longevity: varies

Genus Name (Latin): *Buddleia* Pronunciation: (BUD-lee-uh)

Species Name (Latin): *davidii* cultivar

Common Name: **Butterfly Bush**

Family Name (Latin and Common): *Loganiaceae,* the Logania family

Related Family Members in Book: none

Availability: commercial product: varies June-August

Color: lilac (shown), white, magenta

Unique Characteristics: Plant is fragrant, and has a conical panicle. It is a favorite of butterflies.

Design Applications: (medium) (filler flower) Unique slender flower works well with bold-colored summer flowers.

Longevity: varies

Genus Name (Latin): *Bupleurum* Pronunciation: (bu-PLUR-um)

Species Name (Latin): *rotundifolium*

Common Name: *Bupleurum,* Thoroughwax

Family Name (Latin and Common): *Apiaceae,* the Parsley or Carrot family

Related Family Members in Book: *Ammi, Anethum, Eryngium,* and *Trachymene*

Availability: commercial product: varies year-round

Color: green

Unique Characteristics: Color and form are distinctive.

Design Applications: (medium) (filler flower) Serves as an excellent choice for combining with hot pink and deep purple in vibrant color combinations.

Longevity: days 4-7

Genus Name (Latin): ***Calathea***   Pronunciation: (kal-uh-THEE-uh)

Species Name (Latin): *lancifolia; others*

Common Name: **Rattlesnake**

Family Name (Latin and Common): *Marantaceae,* the Maranta or Arrowroot family

Related Family Members in Book: none

Availability: commercial product: year-round

Color: yellow, green (shown), and white

Unique Characteristics: Flower form is distinctive.

Design Applications: (small) (form flower) Form is best used to add accent; sometimes is useful for line material. Serves as a good choice for contemporary designs, and works best with other exotic or tropical flowers.

Longevity: days 7-10

Genus Name (Latin): ***Calendula***   Pronunciation: (ka-LEN-dew-la)

Species Name (Latin): ***officinalis*** cultivar

Common Name: **Pot Marigold**

Family Name (Latin and Common): *Asteraceae,* the Composite or Sunflower family

Related Family Members in Book: *Achillea, Ageratum, Artemisia, Aster, Centaurea, Chrysanthemum, Cosmos, Craspedia, Dahlia, Echinacea, Echinops, Gerbera, Helianthus, Liatris, Rudbeckia, Senecio, Solidago, x Solidaster,* and *Zinnia*

Availability: commercial product: May-September

Color: orange (shown), yellow, gold, cream, and apricot

Unique Characteristics: popular and durable as a bedding plant for cutting gardens.

Design Applications: (small) (mass flower) Bold color works well in combination with other summer flowers.

Longevity: days 5-7

Genus Name (Latin): ***Callicarpa***   Pronunciation: (kal-i-KAR-pa)

Species Name (Latin): sp.

Common Name: **Beautyberry**

Family Name (Latin and Common): *Verbeniaceae,* the Verbena family

Related Family Members in Book: *Lantana, Verbena*

Availability: Landscape plant has varied commercial availability during September-October as a cut flower

Color: violet-purple

Unique Characteristics: Color of clustered fruit is distinctive.

Design Applications: (small) (form flower) Unique clustered berry adds accent. Berry is excellent choice in combination with autumnal color collections.

Longevity: varies

Genus Name (Latin): ***Callistephus***   Pronunciation: (ka-LIS-te-fus)

Species Name (Latin): *chinensis* 'Matsumoto'

Common Name: **Matsumoto Aster**

Family Name (Latin and Common): *Asteraceae,* the Composite or Sunflower family

Related Family Members in Book: *Achillea, Ageratum, Artemisia, Aster, Centaurea, Chrysanthemum, Cosmos, Craspedia, Dahlia, Echinacea, Gerbera, Helianthus, Liatris, Rudbeckia, Senecio, Solidago, x Solidaster,* and *Zinnia*

Availability: commercial product: June-September

Color: pink, purple

Unique Characteristics: Colors are bold with yellow centers; lateral stems are often long enough to use individually.

Design Applications: (small) (mass flower) Flowers are useful in mixed or bold-colored bouquets, and are a good candidate for paveing technique.

Longevity: days 5-10

Genus Name (Latin): *Callistephus*

Pronunciation: (ka-LIS-te-fus)

Species Name (Latin): *chinensis* cultivar

Common Name: **Mini Matsumoto Aster**

Family Name (Latin and Common): *Asteraceae,* the Composite or Sunflower family

Related Family Members in Book: *Achillea, Ageratum, Artemisia, Aster, Centaurea, Chrysanthemum, Cosmos, Craspedia, Dahlia, Echinacea, Gerbera, Helianthus, Liatris, Rudbeckia, Senecio, Solidago, x Solidaster,* and *Zinnia*

Availability: commercial product: June-September

Color: pink

Unique Characteristics: Color is distinctive. Long lateral stems can be used individually or as a full stem.

Design Applications: (small) (filler flower) Excellent form works well with other bold summer flowers, and is a nice candidate for use in monochromatic color combinations.

Longevity: days 5-10

Genus Name (Latin): *Camellia*

Pronunciation: (ka-MEEL-ya; ka-MEEL-i-a)

Species Name (Latin): spp.

Common Name: *Camellia*

Family Name (Latin and Common): *Theaceae,* the Tea family

Related Family Members in Book: none

Availability: Landscape plant has limited cut flower availability.

Color: white, pink, and red

Unique Characteristics: Fragrance is wonderful.

Design Applications: (small) (mass flower) Flowers can be floated in water, and sometimes used for corsage work.

Longevity: varies

Genus Name (Latin): ***Campanula***   Pronunciation: (kam-PAN-yew-la)

Species Name (Latin): sp.

Common Name: **Canterbury Bells**

Family Name (Latin and Common): *Campanulaceae,* the Bellflower family

Related Family Members in Book: *Trachelium*

Availability: commercial product: April-August

Color: white, light purple (shown)

Unique Characteristics: Flower is bell-shaped.

Design Applications: (medium) (line flower) Depending on the shape of the stem, this flower is useful as a line or filler flower.

Longevity: days 5-7

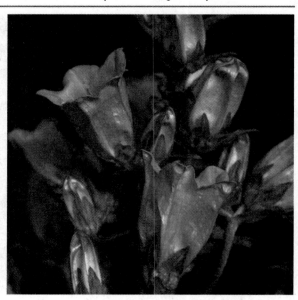

Genus Name (Latin): ***Capsicum***   Pronunciation: (KAP-si-kum)

Species Name (Latin): ***annuum*** cultivar

Common Name: **Ornamental Pepper**

Family Name (Latin and Common): *Solanaceae,* the Nightshade family

Related Family Members in Book: *Brugmansia, Physalis*

Availability: Commercial availability varies year-round as a cut flower and greenhouse plant.

Color: reds (shown), yellows, purple, pink, and green

Unique Characteristics: Species has an assortment of colors and shapes. Fruits change color with stages of maturity.

Design Applications: (medium) (filler flower) This unique plant adds texture, and is useful sometimes as filler. Works well grouped in mass, and also is a good choice for containerized gardens.

Longevity: varies

Genus Name (Latin): ***Carthamus***   Pronunciation: (CAR-tha-mus)

Species Name (Latin): ***tinctorius***

Common Name: **Safflower**, False Saffron

Family Name (Latin and Common): *Asteraceae*, the Composite or Sunflower family

Related Family Members in Book: *Achillea, Ageratum, Artemisia, Aster, Bracteantha, Calendula, Callistephus, Chrysanthemum, Dahlia, Echinacea, Echinops, Gerbera, Helianthus, Liatris, Rudbeckia, Senecio, Tagetes,* and *Zinnia*

Availability: commercial product: June-November

Color: orange

Unique Characteristics: Flower air-dries well upside down, retaining much of its color in both flower and foliage. Lateral stems can be used individually.

Design Applications: (medium) (filler flower) Distinctive flower form provides good texture.

Longevity: weeks 1-2

Genus Name (Latin): ***Catalpa***   Pronunciation: (ka-TAL-pa)

Species Name (Latin): **bignonoides**

Common Name: **Southern Catalpa**, Indian Bean Tree

Family Name (Latin and Common): *Bignoniaceae*, the Bignonia family

Related Family Members in Book: none

Availability: Landscape plant is not available commercially as a cut flower.

Color: white

Unique Characteristics: Flower cluster is large and fragrant.

Design Applications: (large) (mass flower) Adds visual mass to arrangements.

Longevity: varies

Genus Name (Latin): *Catharanthus* syn. *Vinca rosea*   Pronunciation: (ka-tha-RAN-thus)

Species Name (Latin): *roseus* 'Parasol'

Common Name: **Vinca,** Madagascar Periwinkle

Related Family Members in Book: none

Family Name (Latin and Common): *Apocynaceae,* the Dogbane family

Availability: Landscape plant has varied commercial availability.

Color: white (shown) and pinks

Unique Characteristics: Plant is durable.

Design Applications: (small) (mass flower) Provides interesting mass to container gardens, window boxes, and useful as a potted plant for display or staging.

Longevity: varies

Genus Name (Latin): *Cattleya*   Pronunciation: (KAT-lee-a)

Species Name (Latin): many species, hybrids, and cultivars

Common Name: *Cattleya*

Family Name (Latin and Common): *Orchidaceae,* the Orchid family

Related Family Members in Book: *Arachnis, Cymbidium, Dendrobium, Oncidium, Paphiopedilum, Phalaenopsis,* and *Vanda.*

Availability: commercial product: year-round

Color: many

Unique Characteristics: Flower form is distinctive.

Design Applications: (large) (form flower) Blossom is used extensively in bridal work and body flower designs.

Longevity: days 6-10

Genus Name (Latin): ***Celastrus***   Pronunciation: (see-LAS-trus)

Species Name (Latin): ***scandens***

Common Name: **American Bittersweet**

Family Name (Latin and Common): *Celastraceae,* the Staff-Tree family

Related Family Members in Book: *Euonymus*

Availability: commercial product: varies September-October

Color: orange-yellow

Unique Characteristics: Fruit air-dries with slight shrinkage. Capsules open with red seeds; and sometimes are available commercially on long vines as a perishable product or dried material.

Design Applications: (small) (accent) Berry has an interesting coloration and form. Serves as an excellent choice for adding texture and interest to autumnal door decor and everyday designs.

Longevity: long lasting

Genus Name (Latin): ***Celosia***   Pronunciation: (see-LO-shi-a; see-LO-si-a)

Species Name (Latin): ***argentea*** cultivar (Cristata group)

Common Name: **Cockscomb**, Crested Cockscomb

Family Name (Latin and Common): *Amaranthaceae,* the Amaranth family

Related Family Members in Book: *Amaranthus, Gomphrena*

Availability: commercial product: June-October

Color: pink (shown), red, yellow, purple, cerise

Unique Characteristics: Colors are strong and form is distinctive. Plant is often called the "brain" flower. It can be air-dried and other colors are available.

Design Applications: (large) (form and mass flower) Unique form works well in the focal zone with the heavy appearance adding visual mass.

Longevity: days 5-7 or longer

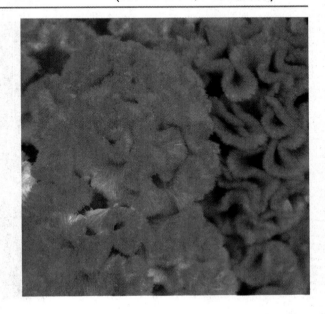

Genus Name (Latin): *Celosia* Pronunciation: (see-LO-shi-a; see-LO-see-a)

Species Name (Latin): *argentea* **'Apricot Brandy'** (Plumosa group)

Common Name: **Cockscomb,** Plume Cockscomb

Family Name (Latin and Common): *Amaranthaceae,* the Amaranth family

Related Family Members in Book: *Amaranthus, Gomphrena*

Availability: commercial product: June-October

Color: orange, red (shown), yellow, pink

Unique Characteristics: Form has feathery plumes. Other colors are available as annual bedding plants and are often bold. Can be air-dried.

Design Applications: (medium) (mass flower) Bold-colored feathery appearance is nice in combination with other summer flowers. Large flower heads add mass.

Longevity: days 5-7 or longer

Genus Name (Latin): *Celosia* Pronunciation: (see-LO-shi-a; see-LO-see-a)

Species Name (Latin): *spicata*

Common Name: **Wheat Celosia**

Family Name (Latin and Common): *Amaranthaceae,* the Amaranth family

Related Family Members in Book: *Amaranthus, Gomphrena*

Availability: commercial product: June-October

Color: pink

Unique Characteristics: Flower has spiked form with white base leading to tip of color. Stems typically contain long, laterals; can be air-dried.

Design Applications: (small) (filler flower) Spike form adds accent to large- and small-scale designs and provides excellent texture.

Longevity: weeks 1-2

Genus Name (Latin): ***Centaurea***

Pronunciation: (sen-ta-REE-ah; sen-TA-ree-ah)

Species Name (Latin): ***cyanus*** cultivar

Common Name: **Bachelor's Button,** Cornflower

Family Name (Latin and Common): *Asteraceae,* the Composite or Sunflower family

Related Family Members in Book: *Achillea, Ageratum, Artemisia, Aster, Chrysanthemum, Cosmos, Craspedia, Dahlia, Echinacea, Echinops, Gerbera, Helianthus, Liatris, Rudbeckia, Senecio, Solidago, x Solidaster,* and *Zinnia*

Availability: commercial product: April-August

Color: blue, pink (shown), and white

Unique Characteristics: The color, form, and size combination of this flower is uncommon among commercial products.

Design Applications: (small) (filler flower) Adorable filler is excellent for garden-style arrangements.

Longevity: days 5-7

Genus Name (Latin): ***Centaurea***

Pronunciation: (sen-ta-REE-ah; sen-TA-ree-ah)

Species Name (Latin): ***macrocephala***

Common Name: **Yellow Cornflower**

Family Name (Latin and Common): *Asteraceae,* the Composite or Sunflower family

Related Family Members in Book: *Achillea, Ageratum, Artemisia, Aster, Chrysanthemum, Cosmos, Craspedia, Dahlia, Echinacea, Echinops, Gerbera, Helianthus, Liatris, Rudbeckia, Senecio, Solidago, x Solidaster,* and *Zinnia*

Availability: commercial product: May-October

Color: yellow

Unique Characteristics: Flower form is distinctive, with a solid globe of color.

Design Applications: (large) (mass flower) Size of flower creates visual weight.

Longevity: days 4-6

Genus Name (Latin): *Cercis*   Pronunciation: (SUR-cis)

Species Name (Latin): *canadensis* **var.** *alba*

Common Name: **Eastern Redbud**

Family Name (Latin and Common): *Fabaceae,* the Pea or Pulse family

Related Family Members in Book: *Acacia, Lathyrus,* and *Wisteria*

Availability: Landscape plant has limited commercial availability as a flowering branch in early spring.

Color: white

Unique Characteristics: Flowers are clustered along leafless stem; white is not as common as the pinkish-purple *C. canadensis.*

Design Applications: (medium) (line flower) Works beautifully in combination with other spring flowers. Branch adds height and length. Serves as a good specimen for Ikebana.

Longevity: varies, depending on bud maturity stage

Genus Name (Latin): *Chaenomeles*   Pronunciation: (kee-NOM-eh-lees)

Species Name (Latin): sp.

Common Name: **Flowering Quince**

Family Name (Latin and Common): *Rosaceae,* the Rose family

Related Family Members in Book: *Cotoneaster, Prunus, Pyrancantha, Pyrus, Rosa,* and *Spiraea.*

Availability: Landscape plant has limited commercial availability as a flowering branch during spring.

Color: orange, apricot (shown), red, pink, and white

Unique Characteristics: Leaf is glossy.

Design Applications: (small) (line flower) Glossy dark green foliage with groups of colored blossoms provide a unique line material for Ikebana.

Longevity: varies

Genus Name (Latin): *Chamelaucium*

Pronunciation: (kam-a-LAW-see-um)

Species Name (Latin): *ciliatum* **'Sterling Range'**

Common Name: **Waxflower**

Family Name (Latin and Common): *Myrtaceae,* the Myrtle family

Related Family Members in Book: *Callistemon, Eucalyptus, Leptospermum, Melaleuca, Myrtus,* and *Thryptomene*

Availability: commercial product: January-May

Color: white

Unique Characteristics: Flower is tiny and massed along linear stems; leaves often appear silvery.

Design Applications: (small) (filler flower) Unique filler can be used as line material when long, lateral stems are available; can provide mass when shorter stems are used.

Longevity: days 7-10

Genus Name (Latin): *Chamelaucium*

Pronunciation: (kam-a-LAW-see-um)

Species Name (Latin): *pheliferum* **'Lady Stephanie'**

Common Name: **Waxflower**

Family Name (Latin and Common): *Myrtaceae,* the Myrtle family

Related Family Members in Book: *Callistemon, Eucalyptus, Leptospermum, Melaleuca, Myrtus,* and *Thryptomene*

Availability: commercial product: September-May

Color: pink

Unique Characteristics: Fragrant, citrus-like smell is apparent upon breaking leaves.

Design Applications: (small) (filler flower) Delicate filler has wonderful light green tips, and is a nice addition to bridal bouquets, corsages, and boutonnieres.

Longevity: days 7-10

Genus Name (Latin): ***Chamelaucium***

Pronunciation: (kam-a-LAW-see-um)

Species Name (Latin): *uncinatum* 'Album'

Common Name: **Waxflower**

Family Name (Latin and Common): *Myrtaceae*, the Myrtle family

Related Family Members in Book: *Callistemon, Eucalyptus, Leptospermum, Melaleuca, Myrtus,* and *Thryptomene*

Availability: commercial product: September–May

Color: white

Unique Characteristics: Fragrant, citrus-like smell is apparent upon breaking leaves.

Design Applications: (small) (filler or line flower) Extremely versatile filler can provide length and height.

Longevity: days 7-10

Genus Name (Latin): ***Chamelaucium***

Pronunciation: (kam-a-LAW-see-um)

Species Name (Latin): *uncinatum* 'Purple Pride'

Common Name: **Waxflower**

Family Name (Latin and Common): *Myrtaceae*, the Myrtle family

Related Family Members in Book: *Callistemon, Eucalyptus, Leptospermum, Melaleuca, Myrtus,* and *Thryptomene*

Availability: commercial product: September–May

Color: purple

Unique Characteristics: Fragrant, citrus-like smell is apparent upon breaking leaves.

Design Applications: (small) (filler or line flower) Extremely versatile filler can be used to add movement to handheld bouquets and provide length or height to container arrangements.

Longevity: days 7-10

Genus Name (Latin): ***Chrysanthemum*** syn. *Dendranthema*   Pronunciation: (kris-AN-thee-mum)

Species Name (Latin): (Indicum group cultivar)

Common Name: **Anemone Pomp,** Duet-Flowered Spray Mum

Family Name (Latin and Common): *Asteraceae,* the Composite or Sunflower family

Related Family Members in Book: *Achillea, Ageratum, Artemisia, Aster, Bracteantha, Centaurea, Cosmos, Craspedia, Dahlia, Echinacea, Echinops, Gerbera, Helianthus, Liatris, Rudbeckia, Senecio, Solidago, x Solidaster,* and *Zinnia*

Availability: commercial product: year-round, many cultivars

Color: white (shown), yellow, bronze

Unique Characteristics: Flower form has raised or cushionlike center.

Design Applications: (small) (mass or filler flower) Individual flower can add mass to any arrangement. Full stem is successful as a filler;

can also be used for paveing or pillowing design techniques.

Longevity: weeks 2-3 long lasting

Genus Name (Latin): ***Chrysanthemum*** syn. *Dendranthema*   Pronunciation: (kris-AN-thee-mum)

Species Name (Latin): (Indicum group cultivar)

Common Name: **Button Pomp,** Button-Flowered Spray Mum

Family Name (Latin and Common): *Asteraceae,* the Composite or Sunflower family

Related Family Members in Book: *Achillea, Ageratum, Artemisia, Aster, Bracteantha, Centaurea, Cosmos, Craspedia, Dahlia, Echinacea, Echinops, Gerbera, Helianthus, Liatris, Rudbeckia, Senecio, Solidago, x Solidaster,* and *Zinnia*

Availability: commercial product: year-round, many cultivars

Color: yellow (shown), white, and green

Unique Characteristics: Flower form has a "button" appearance, and is favored for its longevity and durability.

Design Applications: (small) (mass flower) Round form adds mass. Long, lateral stems can be used individually. Material is useful for paveing or pillowing design techniques.

Longevity: weeks 2-3 long lasting

Genus Name (Latin): ***Chrysanthemum*** syn. *Dendranthema* Pronunciation: (kris-AN-thee-mum)

Species Name (Latin): (Indicum group cultivar)

Common Name: **Cushion Pomp,** Cushion-Flowered Spray Mum

Family Name (Latin and Common): *Asteraceae,* the Composite or Sunflower family

Related Family Members in Book: *Achillea, Ageratum, Artemisia, Aster, Bracteantha, Centaurea, Cosmos, Craspedia, Dahlia, Echinacea, Echinops, Gerbera, Helianthus, Liatris, Rudbeckia, Senecio, Solidago, x Solidaster,* and *Zinnia*

Availability: commercial product: year-round, many cultivars

Color: white (shown), yellow, lavender, purple, bronze, and burgundy

Unique Characteristics: Flower form consists of solid ray florets. Stems often have long laterals. This flower is known for its longevity and durability.

Design Applications: (medium) (mass flower) Round form adds mass to any style of design. Flower is extremely versatile.

Longevity: weeks 2-3 long lasting

Genus Name (Latin): ***Chrysanthemum*** syn. *Dendranthema* Pronunciation: (kris-AN-thee-mum)

Species Name (Latin): (Indicum group cultivar) many cultivars

Common Name: **Daisy Pomp,** Daisy-Flowered Spray Mum

Family Name (Latin and Common): *Asteraceae,* the Composite or Sunflower family

Related Family Members in Book: *Achillea, Ageratum, Artemisia, Aster, Bracteantha, Centaurea, Cosmos, Craspedia, Dahlia, Echinacea, Echinops, Gerbera, Helianthus, Liatris, Rudbeckia, Senecio, Solidago, x Solidaster,* and *Zinnia*

Availability: commercial product: year-round, many cultivars

Color: white (shown), yellow, bronze, burgundy, and lavender

Unique Characteristics: Flower form is the same for all daisy pomps with the center colors varying by cultivar. Stems often have long laterals. This flower is known for its longevity and durability.

Design Applications: (medium) (mass flower) Round form adds mass to any style of design. Flower is extremely versatile.

Longevity: weeks 2-3 long lasting

Genus Name (Latin): ***Chrysanthemum*** syn. *Dendranthema*   Pronunciation: (kris-AN-thee-mum)

Species Name (Latin): (Indicum group cultivar)

Common Name: **Spider Mum,** Fuji

Family Name (Latin and Common): *Asteraceae,* the Composite or Sunflower family

Related Family Members in Book: *Achillea, Ageratum, Artemisia, Aster, Bracteantha, Centaurea, Cosmos, Craspedia, Dahlia, Echinops, Echinacea, Gerbera, Helianthus, Liatris, Rudbeckia, Senecio, Solidago, x Solidaster,* and *Zinnia*

Availability: commercial product: year-round, many cultivars

Color: white, yellow (shown)

Unique Characteristics: 4-5" flowerhead form is produced by many fine to coarse ray florets on a single stem. This is a disbud mum.

Design Applications: (large) (mass flower) Unique "spider" form adds mass and interest; and is useful in traditional line mass designs.

Longevity: weeks 2-3 long lasting

Genus Name (Latin): ***Chrysanthemum*** syn. *Dendranthema*   Pronunciation: (kris-AN-thee-mum)

Species Name (Latin): (Indicum group cultivar)

Common Name: **Spider Pomp,** Spider-Flowered Spray Mum

Family Name (Latin and Common): *Asteraceae,* the Composite or Sunflower family

Related Family Members in Book: *Achillea, Ageratum, Artemisia, Aster, Bracteantha, Calendula, Centaurea, Craspedia, Dahlia, Echinops, Echinacea, Gerbera, Helianthus, Liatris, Rudbeckia, Senecio, Solidago, x Solidaster,* and *Zinnia*

Availability: commercial product: year-round, many cultivars

Color: white, yellow (shown)

Unique Characteristics: Flower form consists of tubular ray florets. Plant is favored for its durability and longevity.

Design Applications: (small) (filler or mass flower) Individual florets can provide interesting filler or mass.

Longevity: weeks 1-3 long lasting

Genus Name (Latin): ***Chrysanthemum*** syn. *Dendranthema*   Pronunciation: (kris-AN-thee-mum)

Species Name (Latin): (Indicum group cultivar)

Common Name: **Football Mum,** Incurve Mum

Family Name (Latin and Common): *Asteraceae,* the Composite or Sunflower family

Related Family Members in Book: *Achillea, Ageratum, Artemisia, Aster, Bracteantha, Centaurea, Cosmos, Craspedia, Dahlia, Echinops, Echinacea, Gerbera, Helianthus, Liatris, Rudbeckia, Senecio, Solidago, x Solidaster,* and *Zinnia*

Availability: commercial product: year-round, many cultivars

Color: white (shown), yellow

Unique Characteristics: Flower form is classified as an intermediate incurve mum. Single-stem flower is favored for its longevity and durability. This is a disbud mum.

Design Applications: (large) (mass flower) Round form adds mass, and is a good candidate for use in traditional line-mass designs.

Longevity: weeks 2-3 long lasting

Genus Name (Latin): ***Chrysanthemum*** syn. *Dendranthema*   Pronunciation: (kris-AN-thee-mum)

Species Name (Latin): (Indicum group cultivar) **'Boris Becker'**

Common Name: **Boris Becker Mum**

Family Name (Latin and Common): *Asteraceae,* the Composite or Sunflower family

Related Family Members in Book: *Achillea, Ageratum, Artemisia, Aster, Bracteantha, Centaurea, Cosmos, Craspedia, Dahlia, Echinops, Echinacea, Gerbera, Helianthus, Liatris, Rudbeckia, Senecio, Solidago, x Solidaster,* and *Zinnia*

Availability: commercial product: year-round, many cultivars

Color: yellow

Unique Characteristics: Flower form consists of tightly nestled ray florets. Flower has a single stem, and leaves appear wider than flower head on stem. This is a disbud mum.

Design Applications: (medium) (mass flower) Unique single stem can be used to add interest or mass.

Longevity: weeks 1-2

Genus Name (Latin): *Chrysanthemum* syn. *Dendranthema* Pronunciation: (kris-AN-thee-mum)

Species Name (Latin): (Indicum group cultivar), **'Tedcha'**

Common Name: **Micro-Pomp,** Spray Mum

Family Name (Latin and Common): *Asteraceae,* the Composite or Sunflower family

Related Family Members in Book: *Achillea, Ageratum, Artemisia, Aster, Bracteantha, Centaurea, Cosmos, Craspedia, Dahlia, Echinops, Echinacea, Gerbera, Helianthus, Liatris, Rudbeckia, Senecio, Solidago, x Solidaster,* and *Zinnia*

Availability: commercial product: year-round, several cultivars

Color: yellow (shown), white, pinks

Unique Characteristics: Most micro-pomps have a flower form typically smaller than other *Chrysanthemum* listings in book. Different colors are available, some with varying center colors.

Design Applications: (small) (filler or mass flower) Short, lateral stem length adapts this pomp to filler and mass applications.

Longevity: weeks 1-3 long lasting

Genus Name (Latin): Chrysanthemum syn. *Dendranthema* Pronunciation: (kris-AN-thee-mum)

Species Name (Latin): (Indicum group cultivar)

Common Name: **Rover,** Red Rover

Family Name (Latin and Common): *Achillea, Asteraceae,* the Composite or Sunflower family

Related Family Members in Book: *Achillea, Ageratum, Artemisia, Aster, Bracteantha, Centaurea, Cosmos, Craspedia, Dahlia, Echinops, Echinacea, Gerbera, Helianthus, Liatris, Rudbeckia, Senecio, Solidago, x Solidaster,* and *Zinnia*

Availability: commercial product: July-November

Color: red tipped yellow

Unique Characteristics: Flower contains yellow tips and is available as a single stem.

Design Applications: (medium) (mass flower) Round form is a good choice for line-mass designs, and a nice choice for autumnal arrangements in combination with other members of the *Chrysanthemum* genus.

Longevity: weeks 1-3 long lasting

Genus Name (Latin): ***Clarkia***   Pronunciation: (CLARK-ee-ah)

Species Name (Latin): ***amoena-*Satin Series** cultivar syn. *Godetia amoena, G. grandiflora*

Common Name: **Godetia,** Satin Flower

Family Name (Latin and Common): *Onagraceae,* the Evening Primrose family

Related Family Members in Book: *Fuchsia*

Availability: commercial product: May-September

Color: reds, pinks, salmon, white, and bicolor

Unique Characteristics: Flower petals have papery texture.

Design Applications: (small) (mass flower) Clustered buds continue to open, adding mass to any design style; can also be used as filler.

Longevity: days 5-7

Genus Name (Latin): ***Clematis***   Pronunciation: (KLEM-a-tis)

Species Name (Latin): **'Jackmanii Alba'**

Common Name: ***Clematis,*** Old Man's Beard, Traveler's Joy, Virgin's Bower

Family Name (Latin and Common): *Ranunculaceae,* the Crowfoot or Buttercup family

Related Family Members in Book: *Aconitum, Anemone, Aquilegia, Consolida, Delphinium, Helleborus, Nigella,* and *Ranunculas*

Availability: Landscape plant is typically not available commercially as a cut flower.

Color: white (shown), pink, purple, yellow, and bicolors

Unique Characteristics: Fragrance in some species; some species are climbers. Sepals range from 4-10 in number. Plant is an interesting accent for interior landscape displays.

Design Applications: (medium) (accent) Provides accent to temporary displays or exhibits.

Longevity: varies as a cut flower

Genus Name (Latin): ***Cleome***   Pronunciation: (klee-OH-mee)

Species Name (Latin): ***hassleriana*** cultivar

Common Name: **Spider Flower**

Family Name (Latin and Common): *Cappari-daceae,* the Caper family

Related Family Members in Book: none

Availability: commercial product: June-August

Color: white, pink, and purple (shown)

Unique Characteristics: Fragrance is distinctive.

Design Applications: (large) (form flower) Unique "spidery" appearance adds interest.

Longevity: days 5-7

Genus Name (Latin): ***Consolida***   Pronunciation: (con-SO-li-da)

Species Name (Latin): ***ajacis*** syn. *C. ambigua, Delphinium consolida*

Common Name: **Larkspur**

Family Name (Latin and Common): *Ranuncu-laceae,* the Crowfoot or Buttercup family

Related Family Members in Book: *Aconitum, Anemone, Aquilegia, Clematis, Delphinium, Helleborus, Nigella,* and *Ranunculas*

Availability: commercial product: June-September

Color: white, pink (shown)

Unique Characteristics: Stems can be hung upside down to air dry. Upon complete drying, florets will remain intact, losing slight color. Feathery foliage is indicative. Flower is ethylene sensitive.

Design Applications: (medium) (line flower) Serves as an excellent choice for adding line, and works especially well in mixed garden bouquets or feminine designs.

Longevity: days 9-10

Genus Name (Latin): **Convallaria**   Pronunciation: (con-vah-LAY-ree-uh)

Species Name (Latin): *majalis*

Common Name: **Lily-of-the-Valley**

Family Name (Latin and Common): *Liliaceae,* the Lily family

Related Family Members in Book: *Aspidistra, Aconitum, Asparagus, Eremurus, Gloriosa, Hemerocallis, Hosta, Kniphofia, Liriope, Muscari, Polygonatum, Ruscus, Smilax, Tulipa,* and *Xerophyllum*

Availability: commercial product: year-round

Color: white

Unique Characteristics: Florets are fragrant and bell-shaped.

Design Applications: (small) (filler flower) Flowers are used primarily for bridal work, and are exceptionally beautiful and fragrant when bundled in mass.

Longevity: days 9-10

Genus Name (Latin): **Coreopsis**   Pronunciation: (ko-ree-OP-sis)

Species Name (Latin): *lanceolata* 'Sunburst'

Common Name: **Tickseed**

Family Name (Latin and Common): *Asteraceae,* the Composite or Sunflower family

Related Family Members in Book: *Achillea, Ageratum, Artemisia, Aster, Centaurea, Chrysanthemum, Cosmos, Craspedia, Dahlia, Echinacea, Echinops, Gerbera, Helianthus, Liatris, Rudbeckia, Senecio, Solidago, x Solidaster,* and *Zinnia*

Availability: Landscape plant is limited commercially as a cut flower.

Color: yellow

Unique Characteristics: Color is intense yellow with daisy appearance on slender stem; typically is available as a perennial plant.

Design Applications: (small) (filler flower) Yellow filler works well in combination with other summer bloomers.

Longevity: varies

Genus Name (Latin): *Coreopsis*

Pronunciation: (ko-ree-OP-sis)

Species Name (Latin): *verticillata* 'Moonbeam'

Common Name: **Tickseed**

Family Name (Latin and Common): *Asteraceae*, the Composite or Sunflower family

Related Family Members in Book: *Achillea, Ageratum, Artemisia, Aster, Bracteantha, Centaurea, Chrysanthemum, Cosmos, Craspedia, Dahlia, Echinacea, Gerbera, Helianthus, Liatris, Rudbeckia, Senecio, Solidago, x Solidaster*, and *Zinnia*

Availability: Landscape plant is limited commercially as a cut flower.

Color: yellow

Unique Characteristics: Foliage is threadlike; typically is available as a perennial plant.

Design Applications: (small) (filler flower) Makes a nice choice for mixed-flower summer bouquets.

Longevity: varies

Genus Name (Latin): *Cornus*

Pronunciation: (KOR-nus)

Species Name (Latin): *florida* 'Cloud Nine'

Common Name: **Dogwood**

Family Name (Latin and Common): *Cornaceae*, the Dogwood family

Related Family Members in Book: *Aucuba*

Availability: Landscape plant is a spring bloomer, but commercially not available as a cut flower.

Color: white

Unique Characteristics: Flower is large on leafless stem.

Design Applications: (medium) (line flower) Distinctive line material immediately resounds spring. Excellent in mass or for Ikebana.

Longevity: varies

Genus Name (Latin): **Cosmos**

Pronunciation: (KOZ-mos; KOZ-mus)

Species Name (Latin): **bipinnatus** cultivar

Common Name: **Cosmos**

Family Name (Latin and Common): *Asteraceae,* the Composite or Sunflower family

Related Family Members in Book: *Achillea, Ageratum, Artemisia, Aster, Bracteantha, Centaurea, Chrysanthemum, Craspedia, Dahlia, Echinacea, Echinops, Gerbera, Helianthus, Liatris, Rudbeckia, Senecio, Solidago, x Solidaster,* and *Zinnia*

Availability: commercial product: June-September

Color: white, pink (shown), and carmine

Unique Characteristics: Foliage is lacy.

Design Applications: (small) (filler flower) Delicate-looking flowers are nice in combination with other summer bloomers; can be used to add mass to smaller-scale designs.

Longevity: days 4-6

Genus Name (Latin): **Costus**

Pronunciation: (KAWS-tus)

Species Name (Latin): **speciosus**

Common Name: **Costus**

Family Name (Latin and Common): *Zingiberaceae,* the Ginger family

Related Family Members in Book: *Alpinia, Curcuma, Globba,* and *Zingiber*

Availability: commercial product: year-round

Color: red

Unique Characteristics: Form is exotic in appearance.

Design Applications: (medium) (form flower) Good choice for use with other exotic flowers; looks best in contemporary designs.

Longevity: weeks 1-2

Genus Name (Latin): ***Craspedia***

Pronunciation: (kra-SPEE-dee-a)

Species Name (Latin): ***uniflora***

Common Name: ***Craspedia,*** Bachelor's Buttons, Billy Buttons

Family Name (Latin and Common): *Asteraceae,* the Composite or Sunflower family

Related Family Members in Book: *Achillea, Ageratum, Artemisia, Aster, Bracteantha, Centaurea, Chrysanthemum, Cosmos, Dahlia, Echinacea, Echinops, Gerbera, Helianthus, Liatris, Rudbeckia, Senecio, Solidago, x Solidaster,* and *Zinnia*

Availability: commercial product: year-round

Color: yellow

Unique Characteristics: Flower air-dries well.

Design Applications: (small) (accent) Drumstick appearance adds interest and movement to contemporary or traditional designs.

Longevity: days 10-14 or longer

Genus Name (Latin): ***Crocosmia***

Pronunciation: (crow-KOZ-mee-ah)

Species Name (Latin): ***x crocosmiiflora***

Common Name: **Montbretia**, Crocosmia

Family Name (Latin and Common): *Iridaceae,* the Iris family

Related Family Members in Book: *Freesia, Gladiolus, Iris, Ixia,* and *Watsonia*

Availability: commercial product: June-October

Color: orange

Unique Characteristics: Stem of flower buds continue to open; and are ethylene sensitive.

Design Applications: (small) (line and form flower) Leafless flower stem with distinctive flower head works well in vegetative and parallel designs. Flower can also provide filler in line mass arrangements. Bold orange works well with complementary colors.

Longevity: days 7-14

Genus Name (Latin): ***Curcuma*** Pronunciation: (kur-KOO-mah)

Species Name (Latin): sp.

Common Name: ***Curcuma,*** Hidden Lily

Family Name (Latin and Common): *Zingiber-aceae,* the Ginger family

Related Family Member in Book: *Alpinia, Costus, Globba,* and *Zingiber*

Availability: commercial product: year-round

Color: pink (shown), yellow

Unique Characteristics: Form is distinctive; bracts are colored.

Design Applications: (small) (form flower) Works best with other tropical flowers, and creates a nice accent in geometric designs.

Longevity: days 5-10

Genus Name (Latin): ***Cyclamen*** Pronunciation: (SIK-la-men; SY-kla-men)

Species Name (Latin): ***persicum*** cultivar

Common Name: **Florist's Cyclamen**

Family Name (Latin and Common): *Primulaceae,* the Primrose family

Related Family Members in Book: *Lysimachia, Primula*

Availability: Indoor plant available year-round.

Color: white (shown), red, salmon, lavender, and pinks

Unique Characteristics: Foliage is heart-shaped; flower is sweet-scented.

Design Applications: (small) (form flower) Distinctive form is a useful choice for corsages or boutonnieres. Blooming plant works well in mixed, containerized gardens.

Longevity: days 2-3 as a cut flower; weeks as a blooming plant

Genus Name (Latin): ***Cymbidium***

Pronunciation: (sim-BID-ee-um)

Species Name (Latin): sp.

Common Name: ***Cymbidium***

Family Name (Latin and Common): *Orchidaceae,* the Orchid family

Related Family Members in Book: *Arachnis, Cattleya, Dendrobium, Oncidium, Paphiopedilum, Phalaenopsis,* and *Vanda*

Availability: commercial product: year-round, many hybrids and cultivars

Color: green, yellow, pink, red, and white

Unique Characteristics: Form is exquisite with stem of buds in full bloom.

Design Applications: (medium) (form and mass flower) When complete stem of florets is used in design, it creates a very elegant and dramatic appearance; it is best used in simple line work. Individually, florets are used in bridal bouquets, corsages, and other body flower creations.

Longevity: weeks 1-3 long lasting

Genus Name (Latin): ***Cynara***

Pronunciation: (SIN-a-ra)

Species Name (Latin): ***scolymus***

Common Name: **Globe Artichoke,** Artichoke

Family Name (Latin and Common): *Asteraceae,* the Composite or Sunflower family

Related Family Members in Book: *Achillea, Ageratum, Artemisia, Aster, Bracteantha, Centaurea, Chrysanthemum, Cosmos, Craspedia, Dahlia, Echinacea, Echinops, Gerbera, Helianthus, Liatris, Rudbeckia, Senecio, Solidago, x Solidaster,* and *Zinnia*

Availability: commercial product: July-October

Color: lavender/purple

Unique Characteristics: Flower has brown-spiked head with tufts of lavender on tip; immature flower heads are cooked as a vegetable.

Design Applications: (large) (form and mass flower) Unique spiked form provides emphasis, and works well in line-mass designs.

Longevity: days 7-10

Genus Name (Latin): *Cytisus* | Pronunciation: (SIT-i-sus)

Species Name (Latin): *scoparius*

Common Name: **Scotch Broom**

Family Name (Latin and Common): *Fabaceae*, the Pea or Pulse family

Related Family Members in Book: *Acacia, Cercis, Lathyrus,* and *Wisteria*

Availability: commercial product: year-round as a foliage, seasonally available with flowers

Color: yellow

Unique Characteristics: Stem is rigid; seasonal flower is a nice addition to dark green stem.

Design Applications: (medium) (line flower) Serves as an excellent choice for adding line.

Longevity: weeks 1-2

Genus Name (Latin): *Cytisus* | Pronunciation: (SIT-i-sus)

Species Name (Latin): **x *spachianus*,** *C. canariensis*

Common Name: **Genista**

Family Name (Latin and Common): *Fabaceae*, the Pea or Pulse family

Related Family Members in Book: *Acacia, Cercis, Lathyrus,* and *Wisteria*

Availability: commercial product: January-May

Color: yellow, white (shown)

Unique Characteristics: Flower is fragrant; leafless stems with multiple florets that continue up the stem.

Design Applications: (small) (filler flower) Arching stem creates soft visual lines; and is a useful filler for adding physical movement to handheld bouquets.

Longevity: days 7-14

Genus Name (Latin): ***Dahlia***

Pronunciation: (DAHL-ya; DAYL-ya)

Species Name (Latin): spp.

Common Name: ***Dahlia***

Family Name (Latin and Common): *Asteraceae,* the Composite or Sunflower family

Related Family Members in Book: *Achillea, Ageratum, Artemisia, Aster, Bracteantha, Centaurea, Cosmos, Chrysanthemum, Craspedia, Echinacea, Echinops, Gerbera, Helianthus, Liatris, Rudbeckia, Senecio, Solidago, x Solidaster,* and *Zinnia*

Availability: commercial product: supply varies, July-November

Color: pink, red, yellow, orange, white, and bicolors

Unique Characteristics: Flower form varies; size varies from buttons to magnificent dinner-plate; many cultivars exist.

Design Applications: (medium) (mass flower) Numerous design applications ranging from traditional to contemporary.

Longevity: days 2-10

Genus Name (Latin): ***Delphinium***

Pronunciation: (del-FIN-ee-um)

Species Name (Latin): **'Yvonne'**—Elatum Group hybrid

Common Name: **Hybrid Delphinium**

Family Name (Latin and Common): *Ranunculaceae,* the Crowfoot or Buttercup family

Related Family Members in Book: *Anemone, Aquilegia, Clematis, Consolida, Helleborus, Nigella,* and *Ranunculas*

Availability: commercial product: year-round

Color: blue shades, white

Unique Characteristics: Color is bold; stems in this group of several hybrids are loose and branched allowing versatility; also is ethylene sensitive.

Design Applications: (medium) (line flower) Works well in garden-style designs, adding intense blue color, and excellent in combination with lime green, orange, or hot pink materials.

Longevity: days 4-12

Genus Name (Latin): *Delphinium*   Pronunciation: (del-FIN-ee-um)

Species Name (Latin): **Pacific Giant** hybrid

Common Name: **Hybrid Delphinium**

Family Name (Latin and Common): *Ranunculaceae,* the Crowfoot or Buttercup family

Related Family Members in Book: *Anemone, Aquilegia, Clematis, Consolida, Helleborus, Nigella,* and *Ranunculas*

Availability: commercial product: year-round

Color: dark blue, lavender-purple (shown), and white

Unique Characteristics: Color is electric blue. Florets are large and typically more compact on stem. Flower is ethylene sensitive. Several hybrids and cultivars exist. Photograph depicts close-up of floret.

Design Applications: (large) (line flower) Linear flower provides mass and intense color, and is beautiful when grouped in volume. Florets can be used individually for body flower designs.

Longevity: days 4-12

Genus Name (Latin): *Dendrobium*   Pronunciation: (den-DRO-bee-um)

Species Name (Latin): sp.

Common Name: **Dendrobium,** Singapore Orchids

Family Name (Latin and Common): *Orchidaceae,* the Orchid family

Related Family Members in Book: *Arachnis, Cattleya, Cymbidium, Oncidium, Paphiopedilum, Phalaenopsis,* and *Vanda*

Availability: commercial product: year-round

Color: yellow, white (shown), lavender, purple, green, and various bicolors

Unique Characteristics: Stems have graceful curve; many hybrids and cultivars are available.

Design Applications: (small) (line flower) Gracefully curved stems add line to traditional or contemporary arrangements; can be used in combination with other tropical flowers. Florets are useful in bridal and body flower designs.

Longevity: days 7-10

Genus Name (Latin): ***Dianthus***   Pronunciation: (die-ANTH-us)

Species Name (Latin): ***caryophyllus*** cultivar

Common Name: **Standard Carnation**

Family Name (Latin and Common): *Caryophyllaceae,* the Pink family

Related Family Members in Book: *Agrostemma, Saponaria*

Availability: commercial product: year-round

Color: white (shown), red, yellow, orange, pink, lavender, purple, green, and various bicolors

Unique Characteristics: Flower is fragrant, long lasting and durable, and extremely versatile. Many cultivars are available. Shows ethylene sensitivity.

Design Applications: (medium) (mass flower) Design uses are unlimited. It remains a staple for the florist due to the variety of color, durability, fragrance, and longevity. Makes an excellent choice for paving, terracing, or pillowing techniques.

Longevity: days 10-14

Genus Name (Latin): ***Dianthus***   Pronunciation: (die-ANTH-us)

Species Name (Latin): ***caryophyllus nana*** cultivar

Common Name: **Mini Carnation,** Pixie Carnation

Family Name (Latin and Common): *Caryophyllaceae,* the Pink family

Related Family Members in Book: *Agrostemma, Saponaria*

Availability: commercial product: year-round

Color: red, yellow, orange, white, pink, lavender, and various bicolors

Unique Characteristics: Flower is fragrant, long lasting, and has many cultivars. Shows ethylene sensitivity.

Design Applications: (small) (mass flower) Unlimited design uses abound for this popular florist staple, from container arrangements to wedding designs. Extremely long-lasting flower and buds add texture, form, and mass.

Longevity: days 10-14

Genus Name (Latin): *Dianthus*

Pronunciation: (die-ANTH-us)

Species Name (Latin): **'Telstar Crimson'**

Common Name: **Pinks,** Chinese Pinks, Annual Pinks

Family Name (Latin and Common): *Caryophyllaceae,* the Pink family

Related Family Members in Book: *Agrostemma, Saponaria*

Availability: annual bedding plant: typically available April-August

Color: red (shown), white, pinks, and bi-colors

Unique Characteristics: Plant is typically used as an annual; other cultivars of pinks are also available as cut flowers. Shows ethylene sensitivity.

Design Applications: (small) (accent) Versatile annual provides color and accent to containerized gardens. Cultivars with long stem length can provide filler to cut flower designs.

Longevity: days 5-9 as a cut flower

Genus Name (Latin): *Dicentra*

Pronunciation: (di-SEN-tra)

Species Name (Latin): *spectabilis*

Common Name: **Bleeding Heart**

Family Name (Latin and Common): *Fumariaceae,* the Fumitory family

Related Family Members in Book: none

Availability: commercial product: supply varies April-May

Color: pink (shown), white

Unique Characteristics: Form of pendulous florets resembles hearts.

Design Applications: (medium) (form flower) Delicate florets are darling by themselves. Nice when bundled in mass.

Longevity: varies, preferring clear water rather than floral foam as water source

Genus Name (Latin): ***Digitalis***

Pronunciation: (dij-i-TAL-is)

Species Name (Latin): ***purpurea***—Excelsior hybrid

Common Name: **Foxglove**

Family Name (Latin and Common): *Scrophulari-aceae,* the Figwort family

Related Family Members in Book: *Antirrhinum, Leucophyllum, Penstemon,* and *Veronica*

Availability: commercial product: June-September

Color: white, yellow, red, and shades of pink

Unique Characteristics: Flower has downward-facing florets with speckled throats; many hybrids and cultivars exist.

Design Applications: (large) (line flower) Useful in formal-linear designs, these flowers are an excellent choice for garden-style arrangements, and a good candidate for vegetative or parallel systems of design.

Longevity: days 7-10

Genus Name (Latin): ***Diosma***

Pronunciation: (die-OZ-ma; die-OS-ma)

Species Name (Latin): ***ericoides***

Common Name: **Diosma,** Coleonema, Breath-of-Heaven

Family Name (Latin and Common): *Rutaceae,* the Rue family.

Related Family Members in Book: *Boronia*

Availability: commercial product: February-May

Color: pink

Unique Characteristics: Fragrant, tiny flower in tight clusters has bold color.

Design Applications: (small) (filler flower) Dainty flower is a delightful filler for romantic styles.

Longevity: days 5-7

Genus Name (Latin): *Diospyros*   Pronunciation: (dy-OHS-py-ros)

Species Name (Latin): sp.

Common Name: **Mini-Persimmon**

Family Name (Latin and Common): *Ebenaceae*, the Ebony family

Related Family Members in Book: none

Availability: commercial product: varies September-October

Color: orange

Unique Characteristics: Fruit grows in small clusters.

Design Applications: (small) (accent) Fruit along a woody stem provides unusual interest. Great addition to autumnal arrangements.

Longevity: varies due to stage of maturity

Genus Name (Latin): *Echinacea*   Pronunciation: (ek-i-NAY-see-a)

Species Name (Latin): *purpurpea* cultivar

Common Name: **Coneflower**

Family Name (Latin and Common): *Asteraceae*, the Composite or Sunflower family

Related Family Members in Book: *Achillea, Ageratum, Artemisia, Aster, Bracteantha, Centaurea, Chrysanthemum, Cosmos, Craspedia, Dahlia, Echinops, Gerbera, Helianthus, Liatris, Rudbeckia, Senecio, Solidago, x Solidaster,* and *Zinnia*

Availability: commercial product: limited June-August

Color: purple, pinkish-purple (shown), and white

Unique Characteristics: Air-dries well, and has a distinctive conelike center.

Design Applications: (large) (mass flower) Daisylike form adds visual mass to mixed summer bouquets.

Longevity: days 5-7

Genus Name (Latin): ***Echinops***

Pronunciation: (EK-I-nops)

Species Name (Latin): *ritro*

Common Name: **Globe Thistle**

Family Name (Latin and Common): *Asteraceae,* the Composite or Sunflower family

Related Family Members in Book: *Achillea, Ageratum, Artemisia, Aster, Bracteantha, Centaurea, Chrysanthemum, Cosmos, Craspedia, Dahlia, Echinacea, Gerbera, Helianthus, Liatris, Rudbeckia, Senecio, Solidago, x Solidaster,* and *Zinnia*

Availability: commercial product: July-October

Color: blue (grey-blue)

Unique Characteristics: Air-dries well, and has metallic blue flowers and thistlelike leaves.

Design Applications: (large) (filler flower) Flower adds a distinctive look to any style design, but is best used in moderation. Makes an excellent choice for contemporary or abstract designs.

Longevity: weeks 1-3 long lasting

Genus Name (Latin): ***Eremurus***

Pronunciation: (er-eh-MEW-rus)

Species Name (Latin): sp.

Common Name: ***Eremurus,*** Desert Candle, Foxtail Lily

Family Name (Latin and Common): *Liliaceae,* the Lily family

Related Family Members in Book: *Asparagus, Aspidistra, Convallaria, Gloriosa, Hemerocallis, Hosta, Hyacinthus, Kniphofia, Lilium, Liriope, Muscari, Ornithogalum, Ruscus, Sandersonia, Tulipa,* and *Xerophyllum*

Availability: commercial product: May-September

Color: yellow (shown), white, cream, pink, and orange

Unique Characteristics: Stem is spikelike with star-shaped florets; colors and availability vary by species and cultivar.

Design Applications: (large) (line flower) Unusual line material for bigger scaled designs.

Longevity: weeks 1-2 with florets continuing to open

Genus Name (Latin): ***Eriostemon***   Pronunciation: (er-i-o-STEE-mon)

Species Name (Latin): sp.

Common Name: ***Eriostemon,*** Waxflower

Family Name (Latin and Common): *Rutaceae,* the Rue family

Related Family Members in Book: *Boronia, Diosma*

Availability: commercial product: April-May

Color: white

Unique Characteristics: Foliage is stiff and waxy looking.

Design Applications: (small) (filler flower) Small flowered filler can also be used as a line element in some designs.

Longevity: days 7-10

Genus Name (Latin): ***Erica***   Pronunciation: (ER-i-ka; ee-RYE-ka)

Species Name (Latin): sp.

Common Name: **Heather**

Family Name (Latin and Common): *Ericaceae,* the Heath family

Related Family Members in Book: *Gaultheria, Rhododendron,* and *Vaccinium*

Availability: commercial product: November-April

Color: pink

Unique Characteristics: Air-dries well. There are several different species, hybrids and/or cultivars are sold commercially.

Design Applications: (small) (filler flower) Dainty flowers along woody stems provide a nice texture, and can sometimes act as a line material; used for bridal designs, corsage work, and container arrangements.

Longevity: weeks 1-2

Genus Name (Latin): *Erica*   Pronunciation: (ER-i-ka; ee-RYE-ka)

Species Name (Latin): *persoluta*

Common Name: **Heather**

Family Name (Latin and Common): *Ericaceae,* the Heath family

Related Family Members in Book: *Gaultheria, Rhododendron,* and *Vaccinium*

Availability: commercial product: November-April

Color: pink

Unique Characteristics: Air-dries well, becoming more fragile and continuing to lose tiny florets. This species of *Erica* is often available as a potted plant.

Design Applications: (small) (filler flower) Tiny, compact florets create an excellent texture. Tips or laterals of stems may also be used for line. Flowers are beautiful in monochromatic color combinations.

Longevity: weeks 1-2

Genus Name (Latin): *Eryngium*   Pronunciation: (eh-RIN-jih-um)

Species Name (Latin): sp.

Common Name: **Sea Holly**

Family Name (Latin and Common): *Apiaceae,* the Parsley or Carrot family

Related Family Members in Book: *Ammi, Anethum, Bupleurum,* and *Trachymene*

Availability: commercial product: June-October

Color: violet, blue, and green (shown)

Unique Characteristics: Air-dries well; form is distinctive; spikey bracts surround cone-shape bracts; color is unique, often with silver or metallic cast; several different species are sold as a commercial cut flower with varying supplies.

Design Applications: (medium) (form flower) This flower with unusual form adds dynamic interest to any design, and is most effective if used in moderation.

Longevity: weeks 1-2

Genus Name (Latin): ***Euphorbia***   Pronunciation: (yew-FOR-bee-a)

Species Name (Latin): ***fulgens*** cultivar

Common Name: **Scarlet-Plume**

Family Name (Latin and Common): *Euphorbiaceae,* the Spurge family

Related Family Members in Book: *Acalypha, Codiaeum* and *Sapium*

Availability: commercial product: September-February

Color: orange, yellow, white, and cream

Unique Characteristics: Stem and leaves release milky sap when cut; condition cut stems in hot water; some people have allergic skin reactions to sap; several cultivars are available.

Design Applications: (medium) (line flower) Tiny florets continue along curving stems providing graceful lines that lengthen designs.

Longevity: days 7-10

Genus Name (Latin): ***Euphorbia***   Pronunciation: (yew-FOR-bee-a)

Species Name (Latin): ***pulcherrima*** cultivar

Common Name: **Poinsettia**, Christmas Flower

Family Name (Latin and Common): *Euphorbiaceae,* the Spurge family

Related Family Members in Book: *Acalypha, Codiaeum* and *Sapium*

Availability: indoor plant: November-January

Color: red (shown), pink, white, yellow, and several mottled and/or speckled varieties

Unique Characteristics: Stem and leaves release milky sap when cut; condition stem in hot water upon removal from plant. Many cultivars are available; plant is sensitive to cool temperatures.

Design Applications: (large) (form flower) This plant is used mostly during the Christmas season. The multiple colors now make it more trans-seasonal.

Longevity: varies as a cut flower

Genus Name (Latin): ***Euphorbia*** Pronunciation: (yew-FOR-bee-a)

Species Name: ***characias*** subsp. ***wulfenii*** **'John Tomlinson'**

Common Name: ***Euphorbia***

Family Name (Latin and Common): *Euphorbiaceae,* the Spurge family

Related Family Members in Book: *Acalypha, Codiaeum,* and *Sapium*

Availability: commercial product: varies from March-May

Color: green (yellow-green)

Unique Characteristics: Stem secretes milky sap; color is unusual.

Design Applications: (medium) (mass) Unique material is an interesting choice for providing mass and accent.

Longevity: days 5-7

Genus Name (Latin): ***Eustoma*** Pronunciation: (you-STO-ma)

Species Name (Latin): ***grandiflorum*** **'Picotee Pink'** syn. *Lisianthus russellianus*

Common Name: **Lisianthus,** Prairie Gentian

Family Name (Latin and Common): *Gentianaceae,* the Gentian family

Related Family Members in Book: *Exacum*

Availability: commercial product: year-round

Color: white, pink, purple, and bicolors (shown)

Unique Characteristics: Flower stem is phototropic, bending toward the light; many cultivars are available. Shows ethylene sensitivity.

Design Applications: (medium) (form flower) Laterals are generally longer and useful in smaller-scale designs. All varieties work well in romantic-style arrangements.

Longevity: days 7-14

Genus Name (Latin): *Exacum*   Pronunciation: (EK-sa-kum)

Species Name (Latin): *affine*

Common Name: *Exacum*

Family Name (Latin and Common): *Gentianaceae,* the Gentian family

Related Family Members in Book: *Eustoma*

Availability: indoor plant: year-round

Color: lavender (shown), pink, and white

Unique Characteristics: Fragrant, saucer-shaped flowers have bright yellow stamens

Design Applications: (small) (mass) This blooming plant works well in mixed gardens of assorted foliage plants. Also a good choice to decorate for individual sales.

Longevity: weeks as a blooming plant

Genus Name (Latin): *Forsythia*   Pronunciation: (for-SITH-ee-ah)

Species Name (Latin): sp.

Common Name: *Forsythia,* Golden-Bells

Family Name (Latin and Common): *Oleaceae,* the Olive family

Related Family Members in Book: *Syringa*

Availability: commercial product: December-March

Color: yellow

Unique Characteristics: Stem is leafless and woody with buds continuing to open.

Design Applications: (large) (line flower) Flowering branch is cheery yellow. Works well for visual displays in early winter, with other forced bulbs, as a welcoming sign of spring. Stems add length and can be bent for use in armatures, or for framing affects.

Longevity: weeks 1-2 or more, depending on stage of maturity upon forcing

Genus Name (Latin): ***Fothergilla***   Pronunciation: (foth-er-GIL-a)

Species Name (Latin): **'Mt. Airy'**

Common Name: **Dwarf Fothergilla**

Family Name (Latin and Common): *Hamamelidaceae*, the Witch Hazel family.

Related Family Members in Book: *Hamamelis*

Availability: landscape plant: limited commercially

Color: white

Unique Characteristics: Flower is fuzzy looking, appearing before leaves.

Design Applications: (small) (form flower) Round-shaped flower with many white filaments provides unique accent.

Longevity: varies

Genus Name (Latin): ***Freesia***   Pronunciation: (FREE-shee-a; FREE-see-a)

Species Name (Latin): ***x hybrida***

Common Name: ***Freesia***

Family Name (Latin and Common): *Iridaceae*, the Iris family

Related Family Members in Book: *Crocosmia, Gladiolus, Iris, Ixia,* and *Watsonia*

Availability: commercial product: year-round

Color: purple, lavender (shown), yellow, white, cream, and orange

Unique Characteristics: Flowers are fragrant appearing on upper side of stem; ethylene sensitivity.

Design Applications: (medium) (form and filler flower) Flower on leafless stem provides a unique form. Works well in contemporary bridal and everyday designs.

Longevity: days 7-10

Genus Name (Latin): *Fuchsia*   Pronunciation: (FEW-sha; FEW-shi-a; FEWK-see-a)

Species Name (Latin): sp.

Common Name: ***Fuchsia,*** Lady's-Eardrops

Family Name (Latin and Common): *Onagraceae,* the Evening Primrose family

Related Family Members in Book: *Clarkia*

Availability: indoor plant: varies commercially

Color: red, pink purple, white, and various bicolors

Unique Characteristics: Flower form is distinctive; color of showy and pendulous florets is dynamic; used as bedding plant, hanging basket, or sometimes shaped into tree forms.

Design Applications: (medium) (form flower) Floret is on trailing stem, and useful in elevated designs where unique form can be visible.

Longevity: varies

Genus Name (Latin): *Gardenia*   Pronunciation: (gar-DEE-nee-a)

Species Name (Latin): sp.

Common Name: ***Gardenia***

Family Name (Latin and Common): *Rubiaceae,* the Madder family

Related Family Members in Book: *Bouvardia, Pentas*

Availability: commercial product: year-round

Color: white

Unique Characteristics: Fragrance is strong. Flowers are commercially available with a waxy-leaf collar attached, easily bruised.

Design Applications: (medium) (form and mass flower) This flower is used extensively for bridal work. It adds both form and visual mass to traditional and contemporary designs.

Longevity: days 1-2

Genus Name (Latin): *Gerbera*

Pronunciation: (ger-BEE-rah; jer-BEE-rah)

Species Name (Latin): *jamesonii* cultivar

Common Name: *Gerbera,* Transvaal Daisy, Barberton Daisy

Family Name (Latin and Common): *Asteraceae,* the Composite or Sunflower family

Related Family Members in Book: *Achillea, Ageratum, Artemisia, Aster, Bracteantha, Centaurea, Chrysanthemum, Cosmos, Craspedia, Dahlia, Echinacea, Echinops, Helianthus, Liatris, Rudbeckia, Senecio, Solidago, x Solidaster,* and *Zinnia*

Availability: commercial product: year-round

Color: yellow, red, orange, pink (shown), and white

Unique Characteristics: Colors are bold and varied; stems are leafless, fluoride sensitivity may cause brown spotting.

Design Applications: (large) (mass flower) Round flower on leafless stem works well in contemporary and traditional designs. Stems

can be wired and shaped, working nicely in abstract or geometric forms.

Longevity: days 4-14, preferring clear water to floral foam

Genus Name (Latin): *Gerbera*

Pronunciation: (ger-BEE-rah; jer-BEE-rah)

Species Name (Latin): *jamesonii* cultivar

Common Name: **Mini Gerbera**

Family Name (Latin and Common): *Asteraceae,* the Composite or Sunflower family

Related Family Members in Book: *Achillea, Ageratum, Artemisia, Aster, Bracteantha, Centaurea, Chrysanthemum, Cosmos, Craspedia, Dahlia, Echinacea, Echinops, Helianthus, Liatris, Rudbeckia, Senecio, Solidago, x Solidaster,* and *Zinnia*

Availability: commercial product: year-round

Color: yellow, pink, orange, and white (shown)

Unique Characteristics: Size is smaller than the traditional *Gerbera;* cultivars are available in varying color; fluoride sensitivity may cause brown spotting.

Design Applications: (small) (mass flower) Flower size works well in smaller-scaled designs.

Longevity: days 5-10, preferring clear water to floral foam

Genus Name (Latin): *Gerbera*   Pronunciation: (ger-BEE-rah; jer-BEE-rah)

Species Name (Latin): *jamesonii* cultivar

Common Name: **Spider Gerbera**

Family Name (Latin and Common): *Asteraceae*, the Composite or Sunflower family

Related Family Members in Book: *Achillea, Ageratum, Artemisia, Aster, Bracteantha, Centaurea, Chrysanthemum, Cosmos, Craspedia, Dahlia, Echinacea, Echinops, Helianthus, Liatris, Rudbeckia, Senecio, Solidago, x Solidaster,* and *Zinnia*

Availability: commercial product: year-round

Color: pink and yellow

Unique Characteristics: Form has a "spider" appearance and is available in standard or mini; fluoride sensitivity may cause brown spotting.

Design Applications: (small) (mass flower) Round form provides mass, and appearance of this variety can certainly be used to add accent.

Longevity: days 5-10, preferring clear water to floral foam.

Genus Name (Latin): *Gladiolus*   Pronunciation: (glad-ee-OH-lus; gla-DEE-oh-lus)

Species Name (Latin): sp.

Common Name: **Gladiolus,** Glad, Sword Lily, Corn Flag

Family Name (Latin and Common): *Iridaceae*, the Iris family

Related Family Members in Book: *Crocosmia, Iris, Ixia,* and *Watsonia*

Availability: commercial product: year-round

Color: white, pink, red, purple, green, yellow, and various bicolors

Unique Characteristics: Flower stem is geotropic, curving away from gravity. Colors are many; florets can be plain, fringed, or ruffled, continuing to open progressively. Photo depicts bud of large, multi-flowered stem.

Design Applications: (large) (line flower) Excellent choice for line-mass designs. Florets are sometimes used in bridal or corsage work to create composite flowers.

Longevity: weeks 1-2

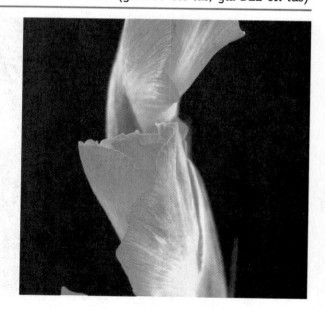

Genus Name (Latin): *Gladiolus*

Pronunciation: (glad-ee-OH-lus; gla-DEE-oh-lus)

Species Name (Latin): *'Nymph'*

Common Name: **Mini Glad,** Mini Gladiolus, Sword Lily, Corn Flag

Family Name (Latin and Common): *Iridaceae,* the Iris family

Related Family Members in Book: *Crocosmia, Iris, Ixia,* and *Watsonia*

Availability: commercial product: varies

Color: white

Unique Characteristics: Stem of florets is loosely arranged on the stem.

Design Applications: (small) (line flower) Flowers are nice when used in mass.

Longevity: weeks 1-2

Genus Name (Latin):*Globba*

Pronunciation: (GLO-bah)

Species Name (Latin): *winitii*

Common Name: *Globba*

Family Name (Latin and Common): *Zingiber-aceae,* the Ginger family

Related Family Members in Book: *Alpinia, Costus, Curcuma,* and *Zingiber*

Availability: commercial product: varies year-round.

Color: pink

Unique Characteristics: Flower bracts are pink; flowers are yellow on pendant racemes; large lance-shaped leaves have a heart-shaped base.

Design Applications: (medium) (form flower) Exotic-looking tropical flower works well in contemporary designs which highlight its form.

Longevity: varies

Genus Name (Latin): *Gloriosa*

Pronunciation: (glow-ree-OH-sa)

Species Name (Latin): *superba* 'Roth-schildiana'

Common Name: **Gloriosa Lily,** Glory Lily, Climbing Lily

Family Name (Latin and Common): *Liliaceae,* the Lily family

Related Family Members in Book: *Asparagus, Aspidistra, Convallaria, Eremurus, Hemerocallis, Hosta, Hyacinthus, Muscari, Ornithagalum, Ruscus, Sandersonia, Smilax, Tulipa,* and *Xerophyllum.*

Availability: commercial product: year-round

Color: yellow to red or purple

Unique Characteristics: Flower form is distinctive with reflexed petals, curled yellow margins, and protruding stamens. Buds continue to open.

Design Applications: (medium) (form flower) Distinctive form adds drama to any design style. It is commonly used in contemporary or abstract designs.

Longevity: days 5-10 per stem

Genus Name (Latin): *Gomphrena*

Pronunciation: (gom-FREE-na)

Species Name (Latin): *globosa* **'Lavender Lady'**

Common Name: *Gomphrena,* Globe Amaranth

Family Name (Latin and Common): *Amaranthaceae,* the Amaranth family

Related Family Members in Book: *Amaranthus, Celosia*

Availability: commercial product: July-September

Color: lavender (shown), white, purple, and red

Unique Characteristics: Air-dries well.

Design Applications: (small) (filler flower) Flowers are generally used as filler.

Longevity: days 5-10

Genus Name (Latin): ***Goniolimon***　　Pronunciation: (go-nee-OH-li-mon)

Species Name (Latin): ***tartarica*** syn. *Limonium tartaricum*

Common Name: **German Statice**

Family Name (Latin and Common): *Plumbaginaceae,* the Plumbago or Leadwort family

Related Family Members in Book: none

Availability: commercial product: year-round

Color: Gray

Unique Characteristics: **Air-dries well, and is prickly.**

Design Applications: (small) (filler flower) Flower is a long-time favorite for drying. Popular in country designs.

Longevity: weeks 2-3 long lasting

Genus Name (Latin): ***Guzmania***　　Pronunciation: (guz-MAY-nee-a)

Species Name (Latin): sp.

Common Name: ***Guzmania***

Family Name (Latin and Common): *Bromeliaceae,* the Bromelia or Pineapple family

Related Family Members in Book: *Aechmea, Ananas,* and *Tillandsia*

Availability: indoor plant: year-round

Color: red (shown), others

Unique Characteristics: Several cultivars are commercially available in varying foliage/flower combinations.

Design Applications: (medium) (accent) This plant is a good choice for adding accent to container gardens and landscape displays.

Longevity: varies

Genus Name (Latin): ***Gypsophila***   Pronunciation: (jip-SOF-i-la)

Species Name (Latin): ***paniculata***

Common Name: **Baby's Breath,** Gyp

Family Name (Latin and Common): *Caryophyllaceae,* the Pink family

Related Family Members in Book: *Agrostemma, Dianthus,* and *Saponaria*

Availability: commercial product: year-round

Color: white

Unique Characteristics: Air-dries well; ethylene sensitivity.

Design Applications: (small) (filler flower) Small flowers provide a delicate appearance. Useful filler has many applications.

Longevity: days 5-7

Genus Name (Latin): ***Gypsophila***   Pronunciation: (jip-SOF-i-la)

Species Name (Latin): **'Million Stars'**

Common Name: **Million Stars Gyp**

Family Name (Latin and Common): *Caryophyllaceae,* the Pink family

Related Family Members in Book: *Agrostemma, Dianthus,* and *Saponaria*

Availability: commercial product: year-round

Color: white

Unique Characteristics: Air dries well; This cultivar has short lateral flower stems; shows ethylene sensitivity.

Design Applications: (small) (filler flower) Small flowers on short stems make this an excellent choice for corsage and boutonniere applications.

Longevity: days 5-7

Genus Name (Latin): *Hamamelis*   Pronunciation: (ham-a-MEE-lis)

Species Name (Latin): sp.

Common Name: **Witch Hazel**

Family Name (Latin and Common): *Hamameli-daceae,* the Witch Hazel family

Related Family Members in Book: *Fothergilla*

Availability: landscape plant: commercially limited as a cut flower

Color: yellow

Unique Characteristics: Flower form is distinctive; blooms appear during late winter.

Design Applications: (small) (accent) Unique flower provides accent.

Longevity: varies

Genus Name (Latin): *Helianthus*   Pronunciation: (hee-li-AN-thus)

Species Name (Latin): sp.

Common Name: **Sunflower**

Family Name (Latin and Common): *Asteraceae,* the Composite or Sunflower family

Related Family Members in Book: *Achillea, Ageratum, Artemisia, Aster, Bracteantha, Centaurea, Chrysanthemum, Cosmos, Craspedia, Dahlia, Echinacea, Echinops, Gerbera, Liatris, Rudbeckia, Senecio, Solidago, x Solidaster,* and *Zinnia*

Availability: commercial product: year-round with peak supplies June-October

Color: yellow

Unique Characteristics: Air-dries well; different cultivars are commercially available with varying sizes.

Design Applications: (large) (mass flower) Round form provides mass. Combines well with summer and autumnal colors.

Longevity: days 7-10

Genus Name (Latin): ***Heliconia***

Species Name (Latin): sp.

Common Name: ***Heliconia,*** Wild Plantain

Family Name (Latin and Common): *Heliconiaceae,* the Heliconia family

Related Family Members in Book: none

Availability: commercial product: year-round

Color: pink (shown), red, yellow, and orange

Unique Characteristics: Unusual form is distinctive.

Design Applications: (large) (form flower) Unique form provides instant drama. Best if used with other tropical flowers. Good choice for geometric designs.

Longevity: days 10-14

Genus Name (Latin): ***Heliconia***

Species Name (Latin): ***pendula*** cultivar

Common Name: **Hanging Heliconia**

Family Name (Latin and Common): *Heliconiaceae,* the Heliconia family

Related Family Members in Book: none

Availability: commercial product: year-round

Color: red-green

Unique Characteristics: Flower form is outstanding; handle with care to keep bracts intact; several cultivars available.

Design Applications: (large) (form flower) This dramatic flower works best when given room to show its unique pendulous habit.

Longevity: days 10-14

Genus Name (Latin): *Heliconia*   Pronunciation: (he-li-KO-ni-a)

Species Name (Latin): *pendula* **'Sexy Pink'**

Common Name: **Hanging Heliconia**

Family Name (Latin and Common): *Heliconiaceae,* the Heliconia family

Related Family Members in Book: none

Availability: commercial product: year-round

Color: pink-green

Unique Characteristics: Form is outstanding; handle with care to keep bracts intact; other cultivars available.

Design Applications: (large) (form flower) This flower is dramatic. Cultivar 'Sexy Pink' is a personal favorite for color and form. Works best when given room to show its unique pendulous habit.

Longevity: days 10-14

Genus Name (Latin): *Heliconia*   Pronunciation: (he-li-KO-ni-a)

Species Name (Latin): *psittacorum* cultivar

Common Name: **Parakeet**

Family Name (Latin and Common): *Heliconiaceae,* the Heliconia family

Related Family Members in Book: none

Availability: commercial product: year-round

Color: pink (shown), orange, and yellow

Unique Characteristics: Flower form is unusual; several cultivars available.

Design Applications: (small) (form flower) Smaller sized *Heliconia* is an excellent choice for tropical designs. This cultivar works well in smaller-scaled arrangements.

Longevity: days 10-14

Genus Name (Latin): *Heliconia*

Pronunciation: (he-li-KO-ni-a)

Species Name (Latin): sp.

Common Name (Latin): *Heliconia*

Family Name (Latin and Common): *Heliconi-aceae,* the Heliconia family

Related Family Members in Book: none

Availability: commercial product: year-round

Color: red

Unique Characteristics: Form is distinctive; several species and cultivars available.

Design Applications: (large) (form flower) Flower's unique form is best if used in combination with other tropical or exotic flowers. Excellent choice for geometric designs.

Longevity: days 10-14

Genus Name (Latin): *Heliotropium*

Pronunciation: (hee-li-oh-TROP-ee-um)

Species Name (Latin): *arborescens* cultivar

Common Name: **Heliotrope**

Family Name (Latin and Common): *Boragi-naceae,* the Borage family

Related Family Members in Book: none

Availability: commercial product: varies

Color: white, purple (shown), blue, and violet

Unique Characteristics: Flower head is dense

Design Applications: (medium) (filler flower) Dense flower head adds mass. A good candidate for pillowing technique.

Longevity: days 4-6

Genus Name (Latin): *Helleborus*

Pronunciation: (he-LEB-oh-rus)

Species Name (Latin): sp.

Common Name: **Christmas Rose,** Lenten Rose, Hellebore

Family Name (Latin and Common): *Ranunculaceae,* the Crowfoot or Buttercup family

Related Family Members in Book: *Anemone, Aquilegia, Clematis, Consolida, Delphinium, Nigella,* and *Ranunculas*

Availability: commercial product: varies year-round

Color: green, pink, purple, and white

Unique Characteristics: Plant is toxic; all parts may cause severe discomfort if ingested; some people may experience skin irritation if in contact with sap; several hybrids and cultivars available commercially.

Design Applications: (medium) (filler flower) This flower is a nice choice for adding interest and mass.

Longevity: days 5-7

Genus Name (Latin): *Hemerocallis*

Pronunciation: (hem-er-oh-KAL-is)

Species Name (Latin): many hybrids and cultivars

Common Name: **Daylily**

Family Name (Latin and Common): *Liliaceae,* the Lily family

Related Family Members in Book: *Asparagus, Aspidistra, Convallaria, Gloriosa, Hosta, Hyacinthus, Kniphofia, Muscari, Ornithogalum, Ruscus, Sandersonia, Smilax, Tulipa,* and *Xerophyllum*

Availability: landscape plant; not commercially available as a cut flower

Color: yellow, red, orange, pink, red, and bicolors

Unique Characteristics: Flower forms are many, with cultivars including triangular, circular, double, star-shaped, and spider-shaped flowers, some with ruffled margins. Blooming period of plant varies by cultivar.

Design Applications: (medium) (form flower) Daylilies provide interest color and accent to temporary or permanent displays.

Longevity: days 1-2 per flower on stem

Genus Name (Latin): ***Heuchera***   Pronunciation: (HOO-ker-a)

Species Name (Latin): ***sanguinea*** cultivar

Common Name: **Coralbells**

Family Name (Latin and Common): *Saxifragaceae,* the Saxifrage family

Related Family Members in Book: *Astilbe*

Availability: commercial product: varies April-May

Color: pink, coral, white, and scarlet red (shown)

Unique Characteristics: Flowers are pendulous, looking like tiny bells along the stem.

Design Applications: (small) (filler flower) Delicate florets along linear stem combine well with other "dainty" flowers.

Longevity: days 3-5

Genus Name (Latin): ***Hibiscus***   Pronunciation: (high-BIS-cus)

Species Name (Latin): sp.

Common Name: ***Hibiscus,*** Mallow, Rose Mallow

Family Name (Latin and Common): *Malvaceae,* the Mallow family

Related Family Members in Book: *Lavatera*

Availability: indoor plant: year-round

Color: yellow, red, pink, white, blue, purple, orange and salmon (shown)

Unique Characteristics: Flower is funnel shaped; several species and cultivars are available.

Design Applications: (large) (accent) Blooming plant adds accent when used with other foliage plants for bridal or party events. Excellent choice for tropical themes.

Longevity: varies

| Genus Name (Latin): ***Hippeastrum*** | Pronunciation: (hip-ee-AS-trum) |

Species Name (Latin): hybrid

Common Name: **Amaryllis,** Barbados Lily

Family Name (Latin and Common): *Amaryllidaceae,* the Amaryllis family

Related Family Members in Book: *Allium, Agapanthus, Narcissus, Nerine,* and *Triteleia*

Availability: commercial product: year-round

Color: red, orange, pink, apricot, salmon, white and bicolors (shown)

Unique Characteristics: Stem is hollow and leafless with four to six trumpet-shaped florets in an umbellate pattern; florets continue to open; some cultivars have faint stripes or colors in petal margin; photo depicts large-flowered cultivar.

Design Applications: (large) (form flower) Showy mass of florets provides drama. Good choice for vegetative or parallel systems of design.

Longevity: varies, depending on maturity stage of buds

| Genus Name (Latin): ***Humulus*** | Pronunciation: (HEW-mew-lus) |

Species Name (Latin): ***lupulus***

Common Name: **Hops**

Family Name (Latin and Common): *Cannabidaceae,* the Hemp family

Related Family Members in Book: none

Availability: commercial product: varies July-August

Color: green

Unique Characteristics: Flower is conelike; available as perishable or preserved and stem-dyed vines.

Design Applications: (texture) Leaf and unique fruit are interesting materials for decorating. Unusual habit works well for arbors, trellises, or garland.

Longevity: days 2-4 as a cut foliage

Genus Name (Latin): ***Hyacinthus***   Pronunciation: (high-a-SIN-thus)

Species Name (Latin): ***orientalis*** cultivar

Common Name: **Hyacinth,** Dutch Hyacinth, Common Hyacinth, Garden Hyacinth

Family Name (Latin and Common): *Liliaceae,* the Lily family

Related Family Members in Book: *Asparagus, Aspidistra, Convallaria, Gloriosa, Hemerocallis, Hosta, Kniphofia, Liriope, Muscari, Ornithagolum, Ruscus, Smilax,* and *Xerophyllum*

Availability: commercial product: November-April

Color: purple (shown), white, blue, red, pink, yellow, and apricot

Unique Characteristics: Fragrance is strong, especially in warmer temperatures

Design Applications: (large) (mass flower) Compact stem of bell-shaped florets adds mass. Works particularly well in combination with other spring flowers in vegetative or parallel

designs. Individual florets can be used for bridal, boutonniere, or corsage work.

Longevity: days 4-7

Genus Name (Latin): ***Hydrangea***   Pronunciation: (hy-DRAN-jee-a)

Species Name (Latin): ***macrophylla*** cultivar

Common Name: **Hydrangea,** Bigleaf Hydrangea

Family Name (Latin and Common): *Hydrangeaceae,* the Hydrangea family

Related Family Members in Book: none

Availability: commercial product: July-October

Color: blue, pink, and white

Unique Characteristics: Air-dries well; color is distinctive with varying shades of blue and pink; displays compound clusters of tiny starlike florets. Several cultivars are available; also available as a potted plant.

Design Applications: (large) (mass flower) Rounded form provides beautiful mass; works well in garden-style arrangements.

Longevity: days 5-7

Genus Name (Latin): **Hydrangea**   Pronunciation: (hy-DRAN-jee-a)

Species Name (Latin): *quercifolia*

Common Name: **Oakleaf Hydrangea**

Family Name (Latin and Common):
Hydrangeaceae, the Hydrangea family

Related Family Members in Book: none

Availability: landscape plant: varies commercially as a cut flower

Color: white

Unique Characteristics: Air-dries well.

Design Applications: (large) (form flower)
Pyramidal-shaped flower head adds mass.

Longevity: days 4-7 as a cut flower

Genus Name (Latin): **Hydrangea**   Pronunciation: (hy-DRAN-jee-a)

Species Name (Latin): *paniculata* 'Tardiva'

Common Name: **Hydrangea Tardiva**

Family Name (Latin and Common):
Hydrangeaceae, the Hydrangea family

Related Family Members in Book: none

Availability: landscape plant: varies commercially as a cut flower

Color: white

Unique Characteristics: Air-dries well.

Design Applications: (large) (mass flower) A
beautiful choice for adding mass; nice textural
addition to monochromatic designs.

Longevity: days 4-7 as a cut flower

Genus Name (Latin): ***Hypericum*** Pronunciation: (hy-PAIR-i-cum)

Species Name (Latin): sp.

Common Name: **St. John's Wort**

Family Name (Latin and Common): *Clusiaceae,* the Garcinia family

Related Family Members in Book: none

Availability: commercial product: varies year-round

Color: green, bronze

Unique Characteristics: Form of capsules is conical; several cultivars with varying shades of green and bronze are available.

Design Applications: (medium) (filler flower) This filler is nice for autumnal designs, and is a good choice for providing texture.

Longevity: days 7-10

Genus Name (Latin): ***Iberis*** Pronunciation: (eye-BEE-ris)

Species Name (Latin): ***umbellata*** Fairy Series

Common Name: **Candytuft**, Rocket Candytuft

Family Name (Latin and Common): *Brassicaceae,* the Mustard family

Related Family Members in Book: *Brassica, Matthiola*

Availability: commercial product: March-August

Color: white (shown), purple, pink, and red

Unique Characteristics: Flower is fragrant sometimes.

Design Applications: (small) (filler flower) Clustered florets work well with summer annuals and mixed bouquets.

Longevity: days 5-7

Genus Name (Latin): *Ilex*

Pronunciation: (EYE-leks)

Species Name (Latin): *verticillata* cultivar

Common Name: **Winterberry,** Common Winter-berry

Family Name (Latin and Common): *Aquifoliaceae,* the Holly family

Related Family Members in Book: none

Availability: commercial product: November-December

Color: red

Unique Characteristics: Fruited (berried) stem is harvested when leaves have dropped from stem; photograph shows foliage intact.

Design Applications: (medium) (line) Medium to large stems of beautiful berries are an excellent choice for adding color. This is a fantastic berry to use in exterior winter plantings and containers as they look dramatic when surrounded by snow. Combines well with conifers in decorations and container arrangements for holiday events.

Longevity: days 7-10 or longer in floral foam; weeks if placed outdoors in cool temperatures

Genus Name (Latin): *Impatiens*

Pronunciation: (im-PAY-shi-enz)

Species Name (Latin): sp.

Common Name: **New Guinea Impatiens**

Family Name (Latin and Common): *Balsaminaceae,* the Balsam or Touch-Me-Not family

Related Family Members in Book: none

Availability: landscape plant: varies

Color: red, orange, pink, purple, and white

Unique Characteristics: Plant is known for its showy foliage; annual flower available in many shades.

Design Applications: (medium) (accent) Works well when massed in landscape beds or used for containerized gardens.

Longevity: varies

Genus Name (Latin): *Impatiens*   **Pronunciation: (im-PAY-shi-enz)**

Species Name (Latin): *wallerana* cultivar

Common Name: *Impatiens,* Busy Lizzy

Family Name (Latin and Common): *Balsami-naceae,* the Balsam or Touch-Me-Not family

Related Members in Book: none

Availability: landscape plant: varies

Color: white (shown), pink, orange, red, violet, and bicolors

Unique Characteristics: Cultivars are many for this shade-loving annual.

Design Applications: (medium) (accent) This favored bedding plant works nicely in containerized gardens as well.

Longevity: varies

Genus Name (Latin): *Impatiens*   **Pronunciation: (im-PAY-shi-enz)**

Species Name (Latin): *wallerana* Rosette Series cultivar

Common Name: Rose Impatiens

Family Name (Latin and Common): *Balsami-naceae,* the Balsam or Touch-Me-Not family

Related Family Members in Book: *Impatiens*

Availability: landscape plant: varies

Color: pink (shown), orange, lavender, white, and red-white bicolor

Unique Characteristics: This shade-loving annual flower has the appearance of a rose.

Design Applications: (medium) (accent) Flower offers unique form.

Longevity: varies

Genus Name (Latin): *Iris*   Pronunciation: (EYE-ris)

Species Name (Latin): many hybrids and cultivars

Common Name: *Iris,* Dutch Iris, Flag

Family Name (Latin and Common): *Iridaceae,* the Iris family

Related Family Members in Book: *Crocosmia, Freesia, Gladiolus, Ixia,* and *Watsonia*

Availability: commercial product: year-round with peak supplies March-May

Color: blue, yellow, purple, and white

Unique Characteristics: Form shown is the florist's variety; will show ethylene sensitivity. Purchase in bud form, tips showing color for best longevity; avoid water loss.

Design Applications: (medium) (form flower) Flower has unique form and is useful in any design style, particularly oriental. Virtually leafless stem can provide strong line.

Longevity: days 2-6

Genus Name (Latin): *Iris*   Pronunciation: (EYE-ris)

Species Name (Latin): many hybrids and cultivars

Common Name: **Bearded Iris,** Flag, Fleur-de-lis

Family Name (Latin and Common): *Iridaceae,* the Iris family

Related Family Members in Book: *Crocosmia, Freesia, Gladiolus, Ixia,* and *Watsonia*

Availability: landscape plant: typically not commercially available as a cut flower

Color: blue, lavender, yellow, white, peach, and bicolors

Unique Characteristics: Form shown is the garden variety; pattern of hairs on the basal half of the falls give this *Iris* a "bearded" appearance; many cultivars exist with varying color combinations in petals, veination, and leaves. Flower is often fragrant.

Design Applications: (large) (form flower) Flowers continue to open along leafless stem of buds; can be used to add mass to mixed summer garden bouquets.

Longevity: days 2-3 for each floret

Genus Name (Latin): *Itea*

Pronunciation: (IT-ee-a)

Species Name (Latin): *virginica* **'Henry's Garnet'**

Common Name: **Virginia Sweetspire**

Family Name (Latin and Common): *Grossulariaceae,* the Currant family

Related Family Members in Book: none

Availability: landscape plant: typically not available as a cut flower

Color: white

Unique Characteristics: Flower is pendulous.

Design Applications: (small) (accent) Small flower with unique pendulous habit is attractive on its own, or to use as accent.

Longevity: days 2-3

Genus Name (Latin): *Ixia*

Pronunciation: (IK-see-a)

Species Name (Latin): sp.

Common Name: ***Ixia,*** Corn Lily, African Corn Lily

Family Name (Latin and Common): *Iridaceae,* the Iris family

Related Family Members in Book: *Crocosmia, Freesia, Gladiolus, Iris,* and *Watsonia*

Availability: commercial product: March-August

Color: pink (shown), orange, yellow, red, and cream

Unique Characteristics: Stem is wiry; florets have varying colored centers.

Design Applications: (small) (line flower) Oblong buds along wiry flower stem add interest and movement. Thin stem allows for use in bud vases with extremely small openings. Nice choice for hand-held bouquets.

Longevity: days 5-10

Genus Name (Latin): ***Kalanchoe***   Pronunciation: (kal-an-KO-ee)

Species Name (Latin): ***blossfeldiana*** cultivar

Common Name: ***Kalanchoe***

Family Name (Latin and Common): *Crassulaceae,* the Orpine family

Related Family Members in Book: *Sedum*

Availability: indoor plant: year-round

Color: red, yellow, orange (shown), pink, salmon, and white

Unique Characteristics: Leaves have a succulent appearance; colors are beautifully bold. Plant is grown as a greenhouse crop; numerous cultivars; should be purchased with slight color showing on buds.

Design Applications: (small) (filler flower) (accent) Florets of intense color add a unique texture to wedding, corsage, and boutonniere designs. Potted plant works well in containerized gardens.

Longevity: days 3-4 as a cut flower

Genus Name (Latin): ***Kniphofia***   Pronunciation: (nip-HO-fee-a; nee-FO-fee-a)

Species Name (Latin): ***uvaria*** hybrid

Common Name: **Red-Hot-Poker,** Tritoma, Torch Lily, Poker Plant

Family Name (Latin and Common): *Liliaceae,* the Lily family.

Related Family Members in Book: *Asparagus, Aspidistra, Convallaria, Gloriosa, Hemerocallis, Hosta, Hyacinthus, Muscari, Ornithagolum, Ruscus, Sandersonia, Smilax,* and *Xerophyllum*

Availability: commercial product: June-October

Color: red, yellow, and orange combinations

Unique Characteristics: Stem of flower is leafless; compact florets give torchlike appearance.

Design Applications: (medium) (line flower) Unique form can add mass and height, and is best suited for contemporary design styles. Allow space to highlight the flower's distinct form and color combination.

Longevity: days 7-10

Genus Name (Latin): ***Kolkwitzia***   Pronunciation: (kol-KWIT-zi-a; kolk-WIT-zi-a)

Species Name (Latin): *amabilis*

Common Name: **Beautybush**

Family Name (Latin and Common): *Caprifoliaceae,* the Honeysuckle family

Related Family Members in Book: *Lonicera, Symphoricarpus, Viburnum,* and *Weigela*

Availability: landscape plant: commercial availability limited as a cut flower

Color: white

Unique Characteristics: Old-fashioned landscape shrub has recent popularity as a cut flower.

Design Applications: (small) (filler flower) Woody stem of small flowers produces a linear filler.

Longevity: varies

Genus Name (Latin): ***Lantana***   Pronunciation: (lan-TAY-na; lan-TAW-na)

Species Name (Latin): *camara* cultivar

Common Name: ***Lantana,*** Shrub Verbena

Family Name (Latin and Common): *Verbeniaceae,* the Vervain or Verbena family

Related Family Members in Book: *Callicarpa, Verbena*

Availability: landscape plant: varies commercially

Color: orange, pink, yellow, cream, and red combinations

Unique Characteristics: Fragrance is strong; typically sold as an annual; several cultivars available.

Design Applications: (small) (accent) Colorful florets are tightly grouped, adding accent. Good choice for containerized plant combinations.

Longevity: varies

Genus Name (Latin): *Lathyrus*   Pronunciation: (LATH-i-rus)

Species Name (Latin): *odoratus* cultivar

Common Name: **Sweet Pea**

Family Name (Latin and Common): *Fabaceae,* the Pea or Pulse family

Related Family Members in Book: *Acacia, Cercis, Cytisus, Lupinus,* and *Wisteria*

Availability: commercial product: February-September

Color: pink, blue, lavender, red, white, and bicolors

Unique Characteristics: Fragrance is very sweet. Garden cut flower has been a favorite for generations; many cultivars exist.

Design Applications: (small) (filler flower) Delicate-looking flower is delightful by itself; nice choice for romantic or feminine design styles.

Longevity: days 3-7

Genus Name (Latin): *Lavatera*   Pronunciation: (la-va-TEE-ra)

Species Name (Latin): *trimestris* cultivar

Common Name: **Tree Mallow**

Family Name (Latin and Common): *Malvaceae,* the Mallow family

Related Family Members in Book: *Hibiscus*

Availability: commercial product: June-October

Color: red, pink, lavender, and white

Unique Characteristics: Flowers appear like miniature *Hibiscus;* several cultivars available with varying shades.

Design Applications: (large) (line flower) Large leafed stem holds flowers and creates a visually heavy line material. Suited for traditional line-mass designs.

Longevity: days 5-7

Genus Name (Latin): *Lavandula*

Pronunciation: (la-VAN-dew-la)

Species Name (Latin): *angustifolia* cultivar

Common Name: **Lavender**

Family Name (Latin and Common): *Lamiaceae,* the Mint family

Related Family Members in Book: *Leonotis, Mentha, Molucella, Monarda, Perovskia, Physostegia, Rosmarinus, Salvia, Solenostemon, Stachys,* and *Thymus*

Availability: commercial product: April-August

Color: lavender, purple

Unique Characteristics: Flower fragrant; air-dries well. Oil of lavender is used in perfumery.

Design Applications: (small) (filler flower) Small, linear flower is wonderful in combination with mixed herb, also works well in parallel or vegetative systems of design.

Longevity: days 3-5

Genus Name (Latin): *Leonotis*

Pronunciation: (lee-o-NO-tis)

Species Name (Latin): *leonurus*

Common Name: **Lion's-Ear**

Family Name (Latin and Common): *Lamiaceae,* the Mint family

Related Family Members in Book: *Lavandula, Mentha, Molucella, Monarda, Perovskia, Physostegia, Rosmarinus, Solenostemon, Stachys,* and *Thymus*

Availability: commercial product: September-October

Color: orange

Unique Characteristics: Flower has whorls of tubular, two-lipped fuzzy florets; stem is square; shows ethylene sensitivity; commercial availability, approximately 6 weeks as a cut flower; plant is a nice choice for large containers in seasonal displays.

Design Applications: (medium) (filler flower) Linear filler plays well against deep purple *Callicarpa* and other bold autumnal colored flowers.

Longevity: varies

Genus Name (Latin): ***Leptospermum***   Pronunciation: (lep-to-SPUR-mum)

Species Name (Latin): ***scoparium*** cultivar

Common Name: ***Leptospermum,*** Lepto

Family Name (Latin and Common): *Myrtaceae,* the Myrtle family

Related Family Members in Book: *Callistemon, Chamelaucium, Eucalyptus, Melaleuca, Myrtus,* and *Thryptomene*

Availability: commercial product: January-April

Color: pink, crimson (shown), and white

Unique Characteristics: Stem is woody; has distinctively coarse texture; air-dries well but tends to shatter easily.

Design Applications: (medium) (filler flower) Nice linear filler provides coarse texture. Full stem and longer laterals can be used as line material.

Longevity: days 7-10

Genus Name (Latin): ***Leucanthemum***   Pronunciation: (loo-KAN-thee-mum)

Species Name (Latin): ***x superbum*** syn. *Chrysanthemum maximum*

Common Name: **Shasta Daisy**

Family Name (Latin and Common): *Asteraceae,* the Composite or Sunflower family

Related Family Members in Book: *Achillea, Ageratum, Artemisia, Aster, Bracteantha, Centaurea, Chrysanthemum, Cosmos, Craspedia, Echinops, Echinacea, Gerbera, Helianthus, Liatris, Rudbeckia, Senecio, Solidago,* x *Solidaster,* and *Zinnia*

Availability: commercial product: varies June-September

Color: white

Unique Characteristics: Foliage has toothed, lance-shaped leaves.

Design Applications: (medium) (mass flower) Round flower provides mass to traditional and contemporary designs. Works well with other summer bloomers in mixed bouquets.

Longevity: days 7-10

Genus Name (Latin): *Leucanthemum*

Pronunciation: (loo-KAN-thee-mum)

Species Name (Latin): ***vulgare*** syn. *Chrysanthemum leucanthemum*

Common Name: **Marguerite Daisy**

Family Name (Latin and Common): *Achillea, Asteraceae,* the Composite or Sunflower family

Related Family Members in Book: *Achillea, Ageratum, Artemisia, Aster, Bracteantha, Centaurea, Chrysanthemum, Cosmos, Craspedia, Dahlia, Echinacea, Echinops, Gerbera, Helianthus, Liatris, Rudbeckia, Senecio, Solidago, x Solidaster,* and *Zinnia*

Availability: commercial product: April-June

Color: white, yellow (shown)

Unique Characteristics: Stem is wiry.

Design Applications: (medium) (mass flower) Cheery flower is darling when bunched in mass. Nice addition to any garden style.

Longevity: days 5-7

Genus Name (Latin): *Leucadendron*

Pronunciation: (loo-ka-DEN-dron)

Species Name (Latin): **'Silvan Red'**

Common Name: **Silvan Red Leucadendron**

Family Name (Latin and Common): *Proteaceae,* the Protea family

Related Family Members in Book: *Banksia, Leucospermum, Protea, Serruria,* and *Telopea*

Availability: commercial product: varies by season

Color: red (burgundy)

Unique Characteristics: Form has exotic appearance.

Design Applications: (medium) (line flower) Linear has exotic appearance. Foliage and flower are useful for adding texture and interest.

Longevity: weeks 2-3 long lasting

Genus Name (Latin): ***Leucadendron***   Pronunciation: (loo-ka-DEN-dron)

Species Name (Latin): **'Safari Sunset'**

Common Name: **Safari Sunset Leucadendron**

Family Name (Latin and Common): *Proteaceae,* the Protea family

Related Family Members in Book: *Banksia, Leucospermum, Protea, Serruria,* and *Telopea*

Availability: commercial product: varies by season

Color: red

Unique Characteristics: Form has exotic appearance.

Design Applications: (medium) (line flower) Material with long laterals makes this a wonderful selection for line elements in exotic and contemporary arrangements; can also provide mass.

Longevity: weeks 2-3 long lasting

Genus Name (Latin): ***Leucospermum***   Pronunciation: (loo-ko-SPUR-mum)

Species Name (Latin): *cordifolium*

Common Name: **Pincushion, Protea**

Family Name (Latin and Common): *Proteaceae,* the Protea family

Related Family Members in Book: *Banksia, Leucadendron, Protea, Serruria,* and *Telopea*

Availability: commercial product: year-round

Color: orange

Unique Characteristics: Flower has unique "pin cushion" appearance; flower spines (styles) can be orange, crimson, or yellow depending on species.

Design Applications: (medium) (mass flower) Mass flower provides accent; great candidate for contemporary or exotic designs.

Longevity: weeks 1-2

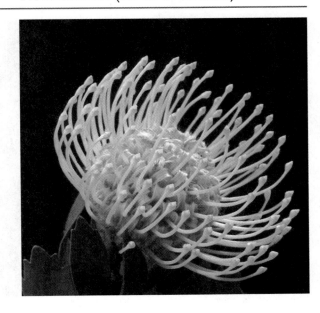

Genus Name (Latin): *Liatris*

Pronunciation: (ly-AY-tris)

Species Name (Latin): *spicata*

Common Name: **Liatris,** Gay-Feather, Blazing-Star

Family Name (Latin and Common): *Asteraceae,* the Composite or Sunflower family

Related Family Members in Book: *Achillea, Ageratum, Artemisia, Aster, Bracteantha, Centaurea, Chrysanthemum, Cosmos, Dahlia, Echinacea, Echinops, Helianthus, Rudbeckia, Senecio, Solidago, x Solidaster,* and *Zinnia*

Availability: commercial product: year-round

Color: purple

Unique Characteristics: Stem of flowers open from the top first, then down.

Design Applications: (medium) (line flower) Line flower is extremely versatile, working especially well in parallel systems of design.

Longevity: days 7-10

Genus Name (Latin): *Lilium*

Pronunciation: (LIL-i-um)

Species Name (Latin): Asiatic hybrid (shown)

Common Name: **Asiatic Lily**

Family Name (Latin and Common): *Liliaceae,* the Lily family

Related Family Members in Book: *Asparagus, Aspidistra, Convallaria, Gloriosa, Hemerocallis, Hosta, Hyacinthus, Kniphofia, Liriope, Muscari, Ornithogalum, Ruscus, Sandersonia, Smilax, Tulipa,* and *Xerophyllum*

Availability: commercial product: year-round

Color: yellow (shown), orange, white, cream, peach, pink, and mauve

Unique Characteristics: Displays a star-shaped form; and petals are often speckled.

Design Applications: (medium) (form flower) Star-shaped flower is a staple for the florist. This flower is known for its versatility in design use, colors, and longevity.

Longevity: days 3-4 per bloom

Genus Name (Latin): *Lilium*   Pronunciation: (LIL-i-um)

Species Name (Latin): *longiflorum*

Common Name: **Easter Lily,** Trumpet Lily

Family Name (Latin and Common): *Liliaceae,* the Lily family

Related Family Members in Book: *Asparagus, Aspidistra, Convallaria, Gloriosa, Hemerocallis, Hosta, Hyacinthus, Kniphofia, Liriope, Muscari, Ornithogalum, Ruscus, Sandersonia, Smilax, Tulipa,* and *Xerophyllum*

Availability: commercial product: year-round

Color: white

Unique Characteristics: Fragrance is strong; flower is trumpet shaped.

Design Applications: (large) (form flower) Flower is beautiful in monochromatic white combinations for religious, sympathy, or bridal events.

Longevity: days 4-5 per bloom

Genus Name (Latin): *Lillium*   Pronunciation: (LIL-i-um)

Species Name (Latin): **'Star Gazer'**

Common Name: **Star Gazer Lily,** Oriental Lily

Family Name (Latin and Common): *Liliaceae,* the Lily family

Related Family Members in Book: *Asparagus, Aspidistra, Convallaria, Gloriosa, Hemerocallis, Hosta, Hyacinthus, Kniphofia, Liriope, Muscari, Ornithogalum, Ruscus, Sandersonia, Smilax, Tulipa,* and *Xerophyllum*

Availability: commercial product: year-round

Color: pink

Unique Characteristics: Flower is fragrant; buds continue to open.

Design Applications: (large) (form flower) Form, size, and fragrance of oriental lilies provide instant accent. They are easily adaptable to any design style from container arrangements to bridal bouquets.

Longevity: days 3-4 per bloom

Genus Name (Latin): *Limonium*

Pronunciation: (ly-MOW-nee-um)

Species Name (Latin): sp.

Common Name: **Sea Lavender**

Family Name (Latin and Common): *Plumbaginaceae,* the Plumbago or Leadwort family

Related Family Members in Book: none

Availability: commercial product: year-round

Color: lavender

Unique Characteristics: Air-dries well with paperlike florets; displays a more open flower head than *L. sinuatum.*

Design Applications: (medium) (filler flower) Form of this filler is useful for adding visual weight. It is also a good candidate for line mass arrangements and the pillowing technique.

Longevity: weeks 1-2

Genus Name (Latin): *Limonium*

Pronunciation: (ly-MOW-nee-um)

Species Name (Latin): *sinuatum* cultivar

Common Name: **Statice**

Family Name (Latin and Common): *Plumbaginaceae,* the Plumbago or Leadwort family

Related Family Members in Book: none

Availability: commercial product: year-round

Color: lavender, purple (shown), blue, yellow, and white

Unique Characteristics: Air-dries well with a papery texture; durable and versatile.

Design Applications: (small) (filler flower) This filler is an excellent choice for adding texture to any style of perishable or dried flower designs.

Longevity: weeks 1-2

Genus Name (Latin): *Limonium*   Pronunciation: (ly-MOW-nee-um)

Species Name (Latin): **Misty Series** cultivar

Common Name: ***Limonium,*** Misty Blue Limonium

Family Name (Latin and Common): *Plumbaginaceae,* the Plumbago or Leadwort family

Related Family Members in Book: *Limonium*

Availability: commercial product: year-round

Color: lavender (shown), others

Unique Characteristics: Air-dries well; appears more open than other *Limoniums.*

Design Applications: (small) (filler flower) Airy filler looks beautiful in bouquets of a monochromatic color scheme.

Longevity: weeks 1-2

Genus Name (Latin): *Limonium*   Pronunciation: (ly-MOW-ni-um)

Species Name (Latin): **'Super Lady'**

Common Name: ***Limonium,*** Super Lady Limonium

Family Name (Latin and Common): *Plumbaginaceae,* the Plumbago or Leadwort family

Related Family Members in Book: *Limonium*

Availability: commercial product: year-round

Color: pink

Unique Characteristics: Air-dries well; florets are paperlike

Design Applications: (small) (filler flower) Filler flower works nicely in feminine designs, and useful for line-mass arrangements; good choice for texture in dried wreaths.

Longevity: weeks 1-2

Genus Name (Latin): *Lobelia*

Pronunciation: (lo-BEE-li-a)

Species Name (Latin): *erinus* cultivar

Common Name: *Lobelia*

Family Name (Latin and Common): *Lobeliaceae*, the Lobelia family

Related Family Members in Book: none

Availability: landscape product: varies commercially

Color: blue, white, purple (shown), and pink-violet

Unique Characteristics: Color is intense; commercially available as an annual; several cultivars sold.

Design Applications: (small) (accent) As a landscape plant this flower provides intense color, looking great in container gardens, especially in combination with complementary colors.

Longevity: varies

Genus Name (Latin): *Lupinus*

Pronunciation: (loo-PIE-nus)

Species Name (Latin): many hybrids and cultivars

Common Name: *Lupine*

Family Name (Latin and Common): *Fabaceae*, the Pea or Pulse family

Related Family Members in Book: *Acacia, Cercis, Cytisus, Lathyrus,* and *Wisteria*

Availability: commercial product: July-September

Color: purple, yellow, mauve, white, carmine, blue

Unique Characteristics: Flowers resemble pea blossoms but on densely packed racemes; several cultivars in various colors are available commercially.

Design Applications: (large) (line flower) Large line material works well with other spring flowers; good choice for line-mass and garden style designs.

Longevity: days 5-10

Genus Name (Latin): *Lysimachia*   Pronunciation: (lie-si-MAY-kie-a; lis-i-MACK-i-a)

Species Name (Latin): *clethroides*

Common Name: **Gooseneck Loosestrife**, Lysimachia, Garden Lysimachia

Family Name (Latin and Common): *Primulaceae,* the Primula family

Related Family Members in Book: *Cyclamen*

Availability: commercial product: July-September

Color: white

Unique Characteristics: Flower has "gooseneck" appearance.

Design Applications: (medium) (filler flower) Flower works well in combination with other summer bloomers, always adding interest by its unique form.

Longevity: days 5-7

Genus Name (Latin): *Magnolia*   Pronunciation: (mag-NO-li-a)

Species Name (Latin): *x soulangiana*

Common Name: **Southern Magnolia**

Family Name (Latin and Common): *Magnoliaceae,* the Magnolia family

Related Family Members in Book: none

Availability: commercial product: limited March-April

Color: white

Unique Characteristics: Form is distinctive with outside of petals flushed pinkish-purple, inside white; flowers appear before the leaves.

Design Applications: (large) (mass flower) Flowers add visual weight. Useful for garden style bouquets.

Longevity: varies

Genus Name (Latin): *Magnolia*　　　Pronunciation: (mag-NO-li-a)

Species Name (Latin): *stellata*

Common Name: **Star Magnolia**

Family Name (Latin and Common): *Magnoli-aceae,* the Magnolia family

Related Family Members in Book: none

Availability: landscape plant: limited availability as a cut flower.

Color: white

Unique Characteristics: Fragrant, star-shaped florets with l2 to 18 narrowly oblong petals; flowers appear before the leaves on stem; large shrub blooms in early spring.

Design Applications: (medium) (accent) Flower adds accent and mass.

Longevity: days 3-5 per bud

Genus Name (Latin): *Malus*　　　Pronunciation: (MAY-lus)

Species Name (Latin): sp.

Common Name: **Flowering Crabapple**

Family Name (Latin and Common): *Rosaceae,* the Rose family

Related Family Members in Book: *Amelanchier, Chaenomeles, Prunus, Pyracantha, Pyrus, Rosa,* and *Spiraea*

Availability: commercial product: varies March-May

Color: white, pink, and red

Unique Characteristics: Stem is woody with single or double florets; several species and cultivars are available in varying shades of color.

Design Applications: (large) (line flower) As in most spring flowering branches, the crabapple provides versatility in line uses. Often a favorite to use in Ikebana designs.

Longevity: days 7-10 or more, depending on bud stage

Genus Name (Latin): *Malus* Pronunciation: (MAY-lus)

Species Name (Latin): sp.

Common Name: **Flowering Crabapple**

Family Name (Latin and Common): *Rosaceae,* the Rose family

Related Family Members in Book: *Amelanchier, Chaenomeles, Prunus, Pyracantha, Pyrus, Rosa,* and *Spiraea*

Availability: commercial product: varies July-September

Color: green shades

Unique Characteristics: Fruit is clustered and shiny.

Design Applications: (small) (filler flower) Wonderful berry provides unique texture.

Longevity: weeks 1-2

Genus Name (Latin): *Matthiola* Pronunciation: (ma-THY-o-la)

Species Name (Latin): *incana* cultivar

Common Name: **Stock,** Gillyflower

Family Name (Latin and Common): *Brassi-caceae,* the Mustard family.

Related Family Members in Book: *Alysum, Brassica,* and *Iberis*

Availability: commercial product: January-October

Color: white, yellowish-cream, lavender (shown), purple, burgundy-wine

Unique Characteristics: Fragrant; sometimes available with roots that are similar to turnip in appearance and scent; many varied colors are available.

Design Applications: (medium) (line flower) This line flower is a staple for the florist. Can also be useful to add mass to varied design styles.

Longevity: days 4-7

Genus Name (Latin): ***Mentha***   Pronunciation: (MEN-tha)

Species Name (Latin): sp.

Common Name: **Mint**

Family Name (Latin and Common): *Lamiaceae,* the Mint family

Related Family Members in Book: *Lavandula, Leonotis, Melissa, Molucella, Monarda, Ocimum, Perovskia, Physostegia, Rosmarinus, Salvia, Solenostemon, Stachys* and *Thymus*

Availability: commercial product: varies

Color: varies by cultivar

Unique Characteristics: fragrant

Design Applications: (small) (filler) Flower and foliage are useful in herbal bouquets.

Longevity: varies

Genus Name (Latin): ***Mertensia***   Pronunciation: (mur-TEN-si-a)

Species Name (Latin): ***virginica***

Common Name: **Bluebells**, Virginia Bluebells, Cowslip, Virginia Cowslip

Family Name (Latin and Common): *Boraginaceae,* the Borage family

Related Family Members in Book: none

Availability: landscape plant: commercially limited as a cut flower

Color: blue and lavender shades

Unique Characteristics: Color is wonderful; flowers are in nodding clusters; prefers clear water to floral foam.

Design Applications: (small) (form flower) Unusual nodding clusters are beautiful alone

Longevity: days 3-4 as a cut flower

Genus Name (Latin): ***Molucella***   Pronunciation: (mo-lyou-SEL-a)

Species Name (Latin): ***laevis***

Common Name: **Bells-of-Ireland,** Molucca Balm, Bellflower

Family Name (Latin and Common): *Lamiaceae,* the Mint family.

Related Family Members in Book: *Lavandula, Leonotis, Mentha, Monarda, Ocimum, Perovskia, Physostegia, Rosmarinus, Salvia, Solenostemon, Stachys,* and *Thymus*

Availability: commercial product: year-round with peak supplies June-October

Color: green

Unique Characteristics: Color is curious green; has shell-like calyces (floral envelopes); flowers are actually tiny, white, and fragrant.

Design Applications: (large) (line flower) Fantastic green color is an essential for St. Patrick's Day events. Excellent choice for line material in solid foliage bouquets. Useful for all design styles.

Longevity: days 7-10

Genus Name (Latin): ***Monarda***   Pronunciation: (mo-NAR-da)

Species Name (Latin): sp.

Common Name: **Bee Balm,** Oswego Tea

Family Name (Latin and Common): *Lamiaceae,* the Mint family

Related Family Members in Book: *Lavandula, Leonotis, Mentha, Molucella, Ocimum, Perovskia, Physostegia, Rosmarinus, Salvia, Solenostemon, Stachys,* and *Thymus*

Availability: commercial product: varies July-October

Color: rose-red, bright crimson, pink, salmon, white, and violet

Unique Characteristics: Fragrant; flower head is whorled.

Design Applications: (medium) (form flower) Flower head is unique and provides interest or mass to various style designs.

Longevity: days 3-5

Genus Name (Latin): *Muscari*

Pronunciation: (mus-KAY-rye)

Species Name (Latin): sp.

Common Name: **Grape Hyacinth**

Family Name (Latin and Common): *Liliaceae,* the Lily family

Related Family Members in Book: *Asparagus, Aspidistra, Convallaria, Gloriosa, Hemerocallis, Hosta, Hyacinthus, Lilium, Liriope, Ornithogalum, Ruscus, Sandersonia, Smilax, Tulipa,* and *Xerophyllum*

Availability: commercial product: January-April

Color: purple-lavender

Unique Characteristics: Flowers are tiny in a dense, short 12 to 20 flowered raceme.

Design Applications: (small) (accent) Flower is best clustered into groups, whether used as a solid mass in a parallel or vegetative design or bundled into a gathering of short glass vases.

Longevity: days 4-7

Genus Name (Latin): *Narcissus*

Pronunciation: (nar-SIS-us)

Species Name (Latin): *tazetta* **'Paper White'**

Common Name: **Paper White Narcissus,** Paper Whites

Family Name (Latin and Common): *Amaryllidaceae,* the Amaryllis family

Related Family Members in Book: *Agapanthus, Allium, Amaryllis, Hippeastrum, Nerine,* and *Triteleia*

Availability: commercial product: November-March

Color: white

Unique Characteristics: Fragrant; often forced into bloom as a potted bulb crop through winter; prefers cooler temperatures to keep stem length shorter.

Design Applications: (small) (mass flower) Clustered flower head is most often used to add mass.

Longevity: days 3-5 as a cut flower; much longer if used as a potted plant

Genus Name (Latin): ***Narcissus***   Pronunciation: (nar-SIS-us)

Species Name (Latin): several hybrids & cultivars

Common Name: **Daffodil,** Jonquil, Narcissus

Family Name (Latin and Common): *Amaryllidaceae,* the Amaryllis family

Related Family Members in Book: *Agapanthus, Allium, Amaryllis, Brodiaea, Hippeastrum, Nerine,* and *Triteleia*

Availability: commercial product: November-April

Color: yellow, white, cream, orange, and bicolors

Unique Charactertistics: Form is unique; single trumpet-shaped flowers are known as daffodils; other varieties are known as jonquils or narcissus; all are part of the genera *Narcissus;* do *not* recut stems when using in combination with other flowers in vase designs; condition alone, since sap secreted when cut is harmful to other flowers.

Design Applications: (medium) (form flower) Trumpet-shaped form adds emphasis and mass to both traditional and contemporary designs.

Longevity: days 4-6

Genus Name (Latin): ***Narcissus***   Pronunciation: (nar-SIS-us)

Species Name (Latin) : **'Tete-a-Tete'**

Common Name: **'Tete-a-Tete'**

Family Name: (Latin and Common): *Amaryllidaceae,* the Amaryllis family

Related Family Members in Book: *Agapanthus, Allium, Amaryllis, Hippeastrum, Nerine,* and *Triteleia*

Availability: indoor plant: January-April

Color: yellow

Unique Characteristics: Flowering bulb plant indicates first sign of spring in many floral shops; other cultivars of similar flower size are periodically available as a cut flower.

Design Applications: (small) (filler flower) Small plant works well in combination with other forced spring bulbs. Makes an excellent choice for containerized gardens.

Longevity: days 3-4 as a cut flower; weeks 1-2 as a potted plant, with florets continuing to open

Genus Name (Latin): *Nerine*

Pronunciation: (nee-RYE-nee; ne-REEN)

Species Name (Latin): *bowdenii*

Common Name: *Nerine,* Guernsey Lily

Family Name (Latin and Common): *Amaryllidaceae,* the Amaryllis family

Related Family Members in Book: *Agapanthus, Allium, Amaryllis, Hippeastrum, Narcissus,* and *Triteleia*

Availability: commercial product: year-round

Color: pink

Unique Characteristics: Form is distinctive; leafless flower stem; funnel-shaped florets in umbel pattern; other species and colors are periodically available.

Design Applications: (small) (form flower). Leafless flower stem works well for parallel or vegetative designs. Unique form adds accent to contemporary-style arrangements. Single florets are useful in corsage or boutonniere work.

Longevity: weeks 1-2

Genus Name (Latin): *Nigella*

Pronunciation: (ny-JEL-a)

Species Name (Latin): *damascena* cultivar

Common Name: **Love-in-a-Mist,** Devil-in-a-Bush

Family Name (Latin and Common): *Ranunculaceae,* the Crowfoot or Buttercup family

Related Family Members in Book: *Aconitum, Anemone, Aquilegia, Clematis, Consolida, Delphinium, Helleborus,* and *Ranunculas*

Availability: commercial product: June-September

Color: blue, pink, and white

Unique Characteristics: Air-dries well; foliage is feathery; many cultivars of this species are commercially available.

Design Applications: (small) (filler flower) Delicate flower has lacy appearance, and provides good texture. Makes an excellent choice for filler.

Longevity: days 5-7

Genus Name (Latin): ***Nymphaea***

Pronunciation: (nim-FEE-a)

Species Name (Latin): sp.

Common Name: **Waterlily**

Family Name (Latin and Common): *Nymphaeaceae,* the Water Lily family

Related Family Members in Book: *Nelumbo*

Availability: commercial product: varies

Color: pink, white, yellow, blue, and red

Unique Characteristics: Fragrant; typically only day blooming varieties are available as a commercial cut flower.

Design Applications: (large) (form flower) Distinctive flower is best when used alone. Good choice for floating designs, for example in glass bubble bowls, or some Ikebana styles.

Longevity: varies

Genus Name (Latin): ***Oncidium***

Pronunciation: (on-SID-i-um)

Species Name (Latin): sp.

Common Name: ***Oncidium,*** Dancing-Lady Orchid, Golden Shower, Dancing Doll

Family Name (Latin and Common): *Orchidaceae,* the Orchid family

Related Family Members in Book: *Arachnis, Cattleya, Cymbidium, Dendrobium, Paphiopedilum, Phalaenopsis,* and *Vanda*

Availability: commercial product: year-round

Color: yellow

Unique Characteristics: Stem of flower is arching with delicately small flowers; provides physical movement in handheld bouquets; photograph shows single floret.

Design Applications: (medium–large) (line flower) Flower stem provides graceful line. Always produces drama and is exceptional when used in combination with other tropicals or exotics.

Longevity: weeks 1-2

Genus Name (Latin): ***Ornithagolum***   Pronunciation: (or-ni-THOG-a-lum)

Species Name (Latin): ***arabicum***

Common Name: **Star-of-Bethlehem**

Family Name (Latin and Common): *Liliaceae,* the Lily family

Related Family Members in Book: *Asparagus, Aspidistra, Convallaria, Gloriosa, Hemerocallis, Hosta, Hyacinthus, Kniphofia, Liriope, Muscari, Polygonatum, Ruscus, Sandersonia, Smilax, Tulipa,* and *Xerophyllum*

Availability: commercial product: year-round

Color: white

Unique Characteristics: Form unique; cup-shaped florets with prominent black center; leafless flower stem.

Design Applications: (medium) (mass flower) Flower head adds mass and accent. Useful as a line material in a number of arrangement styles.

Longevity: weeks 1-2

Genus Name (Latin): ***Ornithagolum***   Pronunciation: (or-ni-THOG-a-lum)

Species Name (Latin): ***thrysoides***

Common Name: **Star-of-Bethlehem,** Chincherinchee

Family Name (Latin and Common): *Liliaceae,* the Lily family

Related Family Members in Book: *Asparagus, Aspidistra, Convallaria, Gloriosa, Hemerocallis, Hosta, Hyacinthus, Kniphofia, Liriope, Muscari, Polygonatum, Ruscus, Sandersonia, Smilax, Tulipa,* and *Xerophyllum*

Availability: commercial product: year-round

Color: white

Unique Characteristics: Form is unique with dense racemes of star-shaped florets; leafless flower stem; long-lasting.

Design Applications: (medium) (mass flower) Flower is versatile. Provides line to parallel systems, mass in mixed bouquets, and always accent to traditional or contemporary styles.

Longevity: weeks 2-3 long lasting

Genus Name (Latin): *Ozothamnus*

Pronunciation: (oh-zo-THAM-nus)

Species Name (Latin): *rosmarinifolius* syn. *Helichrysum rosmarinifolium*

Common Name: **Rice Flower**

Family Name (Latin and Common): *Asteraceae,* the Composite or Sunflower family

Related Family Members in Book: *Achillea, Ageratum, Artemisia, Aster, Bracteantha, Centaurea, Chrysanthemum, Cosmos, Craspedia, Dahlia, Echinacea, Echinops, Gerbera, Helianthus, Liatris, Rudbeckia, Senecio, Solidago, x Solidaster,* and *Zinnia*

Availability: commercial product: varies

Color: white

Unique Characteristics: Fragrant; has dense flower heads; appears reddish in bud, opening to white. Also commercially available as a preserved (glycernized) dried flower.

Design Applications: (small) (mass flower) Flower head is an excellent choice for adding texture and mass. A good candidate for the

pillowing technique. Nice choice for mixed arrangements.

Longevity: weeks 1-2

Genus Name (Latin): *Paeonia*

Pronunciation: (pee-OH-nee-a)

Species Name (Latin): *lactiflora, suffruticosa;* others and cultivars

Common Name: ***Peony***

Family Name (Latin and Common): *Paeoniaceae,* the Peony family

Related Family Members in Book: none

Availability: commercial product: May-July

Color: white, pink, yellow, red, and bicolor

Unique Characteristics: Fragrant; several species and cultivars are available as a commercial cut flower with single, double, and anemone (Japanese) types; Japanese-type shown.

Design Applications: (large) (mass flower) Multipetal blossom visual weight. Makes a beautiful choice for spring bouquets. Great when used in mass.

Longevity: days 3-7

Genus Name (Latin): *Papaver*

Pronunciation: (pah-PAY-ver)

Species Name (Latin): *orientale* 'Allegro'

Common Name: **Poppy,** Oriental Poppy, Iceland Poppy

Family Name (Latin and Common): *Papaveraceae,* the Poppy family

Related Family Members in Book: none

Availability: commercial product: May-July

Color: orange

Unique Characteristics: Flower petals are papery in appearance with a nodding bud and stem tending to be hirsute; petals sometimes have black basal marks; other species and cultivars are commercially available as a cut flower.

Design Applications: (medium) (mass flower) Flower head has a distinctive appearance. Serves as an excellent choice to add interesting mass to garden-style or vegetative designs.

Longevity: days 4-6

Genus Name (Latin): *Papaver*

Pronunciation: (pah-PAY-ver)

Species Name (Latin): spp.

Common Name: **Poppy**

Family Name (Latin and Common): *Papaveraceae,* the Poppy family

Related Family Members in Book: none

Availability: commercial product: summer months

Color: green (light) with gray dusting

Unique Characteristics: Seed head air-dries well, displaying unique color and form.

Design Applications: (small) (accent) Works well in high-style designs, nice element to be grouped and used in parallel systems. Useful for adding interest to mixed summer bouquets.

Longevity: days 5-7 or longer

Genus Name (Latin): ***Paphiopedilum***

Pronunciation: (paf-ee-o-PED-i-lum)

Species Name (Latin): sp.

Common Name: ***Paphiopedilum,*** Lady's Slipper, Slipper Orchid

Family Name (Latin and Common): *Orchidaceae,* the Orchid family

Related Family Members in Book: *Arachnis, Cattleya, Cymbidium, Dendrobium, Oncidium, Phalaenopsis,* and *Vanda*

Availability: commercial product: year-round

Color: green

Unique Characteristics: Form is distinctive; hybrids and cultivars of several colors are available as a commercial cut flower usually with varied striping on sepals and/or lips; greenhouse cypripediums belong to the genera *Paphiopedilum* and *Phragmipedium.* This flower is a personal favorite.

Design Applications: (medium) (form flower) Flower always adds accent and emphasis. Makes

a nice choice for contemporary bridal or everyday designs.

Longevity: days 3-5

Genus Name (Latin): ***Pelargonium***

Pronunciation: (pel-ar-GO-ni-um)

Species Name (Latin): sp.

Common Name: **Scented Geranium**

Family Name (Latin and Common): *Geraniaceae,* the Geranium family

Related Family Members in Book: none

Availability: indoor plant: varies commercially

Color: lavender

Unique Characteristics: Foliage is fragrant, and rubbing leaves makes scent more intense; many assorted cultivars exist with varying fragrances; sometimes ruffled margins appear on leaves. Typically available as a potted plant with occasional availability as a cut flower.

Design Applications: (medium) (form flower) Leaf supports small flower. Due to their unusual shapes and sometimes fuzzy appearance, leaves can be used in bridal work or in combination with other foliages for unique green bouquets.

Longevity: varies as a cut flower and foliage

Genus Name (Latin): *Pelargonium*   Pronunciation: (pel-ar-GO-ni-um)

Species Name (Latin): ***x hortatum*** cultivar

Common Name: **Geranium,** Zonal Geranium

Family Name (Latin and Common): *Geraniaceae,* the Geranium family

Related Family Members in Book: none

Availability: landscape plant: varies commercially

Color: red, pink, white, salmon, scarlet and orange

Unique Characteristics: Color shades are varied; this annual has many cultivars, including single, double, or cactus-flowered and fancy-leaved varieties.

Design Applications: (medium) (mass flower) Flower is available in numerous colors. Makes a good choice for mixed containerized plantings.

Longevity: varies as bedding plant, short-lived as a cut flower

Genus Name (Latin): *Pentas*   Pronunciation: (PEN-tas)

Species Name (Latin): ***lanceolata* 'Pink Profusion'**

Common Name: **Egyptian Star Cluster,** Star Cluster, Pentas

Family Name (Latin and Common): *Rubiaceae,* the Madder family

Related Family Members in Book: *Bouvardia, Gardenia*

Availability: commercial product: July-September

Color: pink

Unique Characteristics: Form consists of flat or domed corymbs of flowers; several colors are available; cultivar shown is typically available as a bedding plant and shorter than others sold as commercial cut flower.

Design Applications: (medium) (form flower) This flower is a good choice for basal focal zones. Also used to provide mass in vase or container arrangements of mixed summer flowers.

Longevity: days 4-7 as a cut flower

Genus Name (Latin): ***Perovskia***   Pronunciation: (pe-ROF-ski-ya)

Species Name (Latin): sp.

Common Name: **Russian Sage**

Family Name (Latin and Common): *Lamiaceae,* the Mint family

Related Family Members in Book: *Lavandula, Leonotis, Mentha, Monarda, Ocimum, Perovskia, Physostegia, Rosmarinus, Salvia, Solenostemon, Stachys,* and *Thymus*

Availability: commercial product: varies July-September

Color: lavender, blue

Unique Characteristics: Foliage is fragrant.

Design Applications: (small) (filler flower) Plant has wonderful gray-green foliage providing good visual support for blue-violet flowers. Works well as a filler in garden-style bouquets.

Longevity: days 3-5

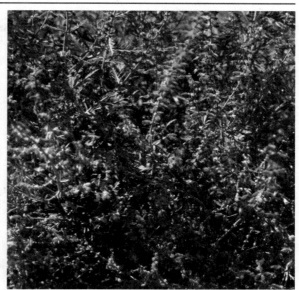

Genus Name (Latin): ***Phalaenopsis***   Pronunciation: (fal-eh-NOP-sis)

Species Name (Latin): spp.

Common Name: **Moth Orchid**

Family Name (Latin and Common): *Orchidaceae,* the Orchid family

Related Family Members in Book: *Arachnis, Cattleya, Cymbidium, Dendrobium, Oncidium, Paphiopedilum,* and *Vanda*

Availability: commercial product: year-round

Color: white

Unique Characteristics: Form is distinctive, having the appearance of a moth; cut flowers are sold in sprays or singly; requires special wiring techniques when used individually; several hybrids are commercially available, typically in white or pink, often having colored lips.

Design Applications: (medium) (form flower) Flower is used extensively in bridal and corsage work. Full sprays of florets are exquisite.

Longevity: days 3-5

Genus Name (Latin): **Phlox**   Pronunciation: (FLOKS)

Species Name (Latin): **paniculata** cultivar

Common Name: **Garden Phlox**

Family Name (Latin and Common): *Polemoniaceae,* the Phlox family

Related Family Members in Book: none

Availability: commercial product: June-October

Color: lavender, pink, red, white, and bicolor

Unique Characteristics: Fragrant; flower head has small florets in dense terminal panicles; several cultivars of species are available in varying shades of color.

Design Applications: (large) (mass flower) Dense flower head works well in mixed summer bouquets.

Longevity: days 3-7

Genus Name (Latin): **Phylica**   Pronunciation: (fy-LEE-ka)

Species Name (Latin): sp.

Common Name: **Phylica,** Cape Myrtle

Family Name (Latin and Common): *Rhamnaceae,* the Buckthorn family

Related Family Members in Book: none

Availability: commercial product: varies

Color: yellow (pale)

Unique Characteristics: Flower head is hairy or downy in appearance.

Design Applications: (small) (filler flower) Fuzzy-looking filler flower is a good choice for adding interest and texture.

Longevity: days 7-10

Genus Name (Latin): ***Physalis***   Pronunciation: (FY-sa-lis; FIS-a-lis)

Species Name (Latin): *alkekengi*

Common Name: **Chinese Lantern**, Japanese Lantern, Ground Cherry

Family Name (Latin and Common): *Solanaceae*, the Nightshade family

Related Family Members in Book: *Capsicum*

Availability: commercial product: August-October

Color: orange

Unique Characteristics: Air-dries well; red orange berries are enclosed in papery calyces; individual lanterns are 1–2 inches in size; typically available as shown, without leaves.

Design Applications: (small) (form flower) Distincitive lanterns are a good choice for exotic autumnal bouquets. Single florets can add accent. Dramatic stem can be used as a line element.

Longevity: varies

Genus Name (Latin): ***Physostegia***   Pronunciation: (fy-so-STEE-ji-a)

Species Name (Latin): *virginiana* cultivar

Common Name: ***Physostegia,*** False Dragonhead, Obedient Plant

Family Name (Latin and Common): *Lamiaceae*, the Mint family

Related Family Members in Book: *Lavandula, Leonotis, Mentha, Molucella, Monarda, Ocimum, Perovskia, Rosmarinus, Salvia, Solenostemon, Stachys,* and *Thymus*

Availability: commercial product: July-October

Color: pinks, lavender, and white

Unique Characteristics: Stem is square; leaves are toothed; several cultivars are available commercially.

Design Applications: (small) (filler flower) Slender form is a good filler to mixed bouquets, adding unique interest.

Longevity: days 5-7

Genus Name (Latin): ***Polianthes***

Pronunciation: (pol-i-AN-theez)

Species Name (Latin): ***tuberosa***

Common Name: **Tuberose**

Family Name (Latin and Common): *Agavaceae,* the Agave family

Related Family Members in Book: *Cordyline, Dracaena, Phormium, Yucca*

Availability: commercial product: February-October

Color: white

Unique Characteristics: Fragrance is strong; flower is tubular, with waxy-appearance.

Design Applications: (large) (line flower) Multi-flowered stem is an excellent choice for line-mass designs. Florets are often useful in corsages and boutonnieres.

Longevity: weeks 1-2

Genus Name (Latin): ***Portulaca***

Pronunciation: (por-TYOU-lay-ka; por-tyou-LAK-a)

Species Name (Latin): ***grandiflora*** hybrid

Common Name: **Rose Moss,** Purslane

Family Name (Latin and Common): *Portula-caceae,* the Purslane family

Related Family Members in Book: none

Availability: landscape plant: varies commercially

Color: pink, fuchsia, yellow, cream, scarlet, apricot, and white

Unique Characteristics: Colors are bold; profuse bloomer of double and single flowers; typically available as an annual; low light levels inhibit flowers from opening.

Design Applications: (small) (texture) Vivid flower appears on stem of succulent-looking foliage, making this an interesting addition of texture to containerized plantings.

Longevity: varies

Genus Name (Latin): ***Protea*** Pronunciation: (PRO-tee-a)

Species Name (Latin): ***cynaroides***

Common Name: **King Protea**

Family Name (Latin and Common): *Proteaceae,* the Protea family

Related Family Members in Book: *Banksia, Leucadendron, Leucospermum, Serruria,* and *Telopea*

Availability: commercial product: year-round

Color: pink

Unique Characteristics: Form is distinctive; color is in bract that is silky/hairy on the outside; large, almost 8-inch flower head, readily distinguishes this species; color ranges from light pink to dark pink or nearly red.

Design Applications: (large) (form flower) Dramatic flower demands attention. Excellent in combination with other exotics.

Longevity: weeks 2-3 long lasting

Genus Name (Latin): ***Protea*** Pronunciation: (PRO-tee-a)

Species Name (Latin): ***eximia***

Common Name: **Rose Spoon Protea,** Duchess

Family Name (Latin and Common): *Proteaceae,* the Protea family

Related Family Members in Book: *Banksia, Leucadendron, Leucospermum, Serruria,* and *Telopea*

Availability: commercial product: year-round

Color: red or red-tinted pink

Unique Characteristics: Form is distinctive; colorful bracts are fringed with white hair; can be air-dried.

Design Applications: (medium–large) (form flower) Like most *Proteas,* this inflorescence is distinctive and works best placed in focal points or zones.

Longevity: weeks 2-3 long lasting

Genus Name (Latin): ***Protea***

Pronunciation: (PRO-tee-a)

Species Name (Latin): ***nerifolia*** cultivar

Common Name: **White Mink Protea**

Family Name (Latin and Common): *Proteaceae,* the Protea family

Related Family Members in Book: *Banksia, Leucadendron, Leucospermum, Serruria,* and *Telopea*

Availability: commercial product: year-round

Color: white

Unique Characteristics: Form is distinctive; bracts are fringed with black hairs; pink cultivar is also commonly used; can be air-dried.

Design Applications: (medium–large) (form flower) Flower has distinctive form, always adding interest and mass to any design. Flower is best used in focal zones, and is well suited for contemporary styles.

Longevity: weeks 2-3 long lasting

Genus Name (Latin): ***Protea***

Pronunciation: (PRO-tee-a)

Species Name (Latin): ***repens***

Common Name: **Honeyflower,** Sugarbush

Family Name (Latin and Common): *Proteaceae,* the Protea family

Related Family Members in Book: *Banksia, Leucadendron, Leuspermum, Serruria,* and *Telopea*

Availability: commercial product: year-round

Color: white

Unique Characteristics: Form is unique with bracts being covered by sticky resin; bracts are often tipped with dark red or pink.

Design Applications: (small-medium) (form flower) Small *Protea* provides accent as well as mass. Works uniquely well in combination with other exotic materials.

Longevity: weeks 2-3 long lasting

Genus Name (Latin): ***Prunus***   Pronunciation: (PROO-nus)

Species Name (Latin): sp.

Common Name: **Flowering Cherry**

Family Name (Latin and Common): *Rosaceae,* the Rose family

Related Family Members in Book: *Amelanchier, Chaenomeles, Malus, Pyracantha, Pyrus, Rosa,* and *Spiraea*

Availability: commercial product: March-April

Color: pink

Unique Characteristics: Stem is woody with flower typically appearing before leaves; genus includes plum, cherry, peach, apricot and nectarine; photo shows single petaled form.

Design Applications: (large) (line flower) Spring flowering stem provides beautiful line element for any design style. Wonderful when used in mass for dramatic window displays. Makes a good choice for Ikebana.

Longevity: weeks 1-2

Genus Name (Latin): ***Prunus***   Pronunciation: (PROO-nus)

Species Name (Latin): sp.

Common Name: **Flowering Cherry**

Family Name (Latin and Common): *Rosaceae,* the Rose family

Related Family Members in Book: *Amelanchier, Chaenomeles, Malus, Pyracantha, Pyrus, Rosa,* and *Spiraea*

Availability: commercial product: March-April

Color: pink

Unique Characteristics: Stem is woody with delicate pink or white flowers; genus includes plum, cherry, peach, apricot, almond, and nectarine; photo shows double petaled ornamental variety.

Design Applications: (large) (line flower) Flowering branch works well in combination with other spring flowers. Dramatic when used in mass. All *Prunus* work especially well in Oriental designs.

Longevity: weeks 1-2

Genus Name (Latin): *Pyracantha*   Pronunciation: (py-ra-KAN-tha)

Species Name (Latin): spp.

Common Name: ***Pyracantha***

Family Name (Latin and Common): *Rosaceae,* the Rose family

Related Family Members in Book: *Amelanchier, Chaenomeles, Malus, Prunus, Pyrus, Rosa,* and *Spiraea*

Availability: landscape plant: varies commercially, August-October as a cut berry

Color: orange

Unique Characteristics: Periodically this thorny plant appears with its beautiful berries as a commercial-cut product.

Design Applications: (large) (mass) Clusters of berries among lustrous dark green leaves are a nice addition to designs for color, accent, texture, or mass.

Longevity: days 7-10

Genus Name (Latin): *Pyrus*   Pronunciation: (PIE-rus)

Species Name (Latin): *calleryana* **'Bradford'**

Common Name: **Bradford Callery Pear**

Family Name (Latin and Common): *Rosaceae,* the Rose family

Related Family Members in Book: *Amelanchier, Chaenomeles, Malus, Prunus, Pyracantha, Rosa,* and *Spiraea*

Availability: landscape plant: typically not available as a commercial cut flower

Color: white

Unique Characteristics: Stem is woody with early spring flowering tree producing clusters of white flowers; has a slight fragrance that some find offensive.

Design Applications: (large) (line flower) Branching plant can be a good choice for Ikebana.

Longevity: varies

Genus Name (Latin): *Ranunculas*

Pronunciation: (ra-NUNG-kew-lus)

Species Name (Latin): spp.

Common Name: **Ranunculas,** Persian Buttercup

Family Name (Latin and Common): *Ranunculaceae,* the Crowfoot or Buttercup family

Related Family Members in Book: *Aconitum, Anemone, Aquilegia, Clematis, Consolida, Delphinium, Helleborus,* and *Nigella*

Availability: commercial product: January-May

Color: pink, red, orange (shown), yellow, and white

Unique Characteristics: Color is bold with contrasting center; double flowered form is most readily available in the cut flower trade; other hybrids and cultivars periodically appear for purchase.

Design Applications: (medium) (mass flower) Round flower head works well in any styles of design, traditional or contemporary. Buds and leaves can also be used to add texture or interest.

Longevity: days 4-7

Genus Name (Latin): *Rhododendron* syn *Azalea*

Pronunciation: (row-doe-DEN-dron)

Species Name (Latin): spp.

Common Name: *Rhododendron*

Family Name (Latin and Common): *Ericaceae,* the Heather family

Related Family Members in Book: *Erica, Gaultheria,* and *Vaccinium*

Availability: landscape plant: varies commercially as a cut flower

Color: red, pink, white, cream, lavender (shown), and orange

Unique Characteristics: Colors are many; glossy green foliage on broad-leaf evergreen; leaves are generally larger and broader than what is commonly referred to as an azalea among florists.

Design Applications: (medium–large) (accent) Landscape plant is an excellent addition to temporary landscape displays or exhibits.

Longevity: varies

Genus Name (Latin): ***Rhododendron*** syn. *Azalea*   Pronunciation: (row-doe-DEN-dron)

Species Name (Latin): spp.

Common Name: **Florist's Azalea**

Family Name (Latin and Common): *Ericaceae,* the Heath family

Related Family Members in Book: *Erica, Gaultheria,* and *Vaccinium*

Availability: indoor plant: year-round

Color: white, pink, salmon, crimson, and bicolors

Unique Characteristics: Colors are many; florists' azaleas are typically grown as a greenhouse crop.

Design Applications: (small–medium) (color) Blooming plant is a good candidate for container gardens and temporary displays. Florets can be used in corsages.

Longevity: varies

Genus Name (Latin): ***Rhododendron*** syn. *Azalea*   Pronunciation: (row-doe-DEN-dron)

Species Name (Latin): spp.

Common Name: ***Azalea***

Family Name (Latin and Common): *Ericaceae,* the Heath family

Related Family Members in Book: *Erica, Gaultheria,* and *Vaccinium*

Availability: landscape plant: commercial availability limited as a cut flower.

Color: red, pink (shown), white, cream, lavender, apricot, yellow, and orange

Unique Characteristics: Colors are many for this favored landscape plant; bloom times vary depending on cultivar; even though a *Rhododendron* by genus, the smaller leaf species are often referred to as azaleas.

Design Applications: (medium–large) (color) Blooming plant is useful for temporary exhibits or displays. Makes an excellent addition to spring vignettes.

Longevity: varies

Genus Name (Latin): ***Rosa***   Pronunciation: (RO-sa)

Species Name (Latin): *floribunda, grandiflora, rugosa;* others

Common Name: **Rose Hips**

Family Name (Latin and Common): *Rosaceae,* the Rose family

Related Family Members in Book: *Amelanchier, Chaenomeles, Malus, Prunus, Pyracantha, Pyrus,* and *Spiraea*

Availability: commercial product: September-November

Color: orange to red shades

Unique Characteristics: Form is wonderful; contains thorny stem. Can be air-dried, shrinking slightly in size and eventually dropping.

Design Applications: (small) (color) Fruit is a great choice for texture in mixed autumnal bouquets.

Longevity: weeks 1-2 or longer

Genus Name (Latin): ***Rosa***   Pronunciation: (RO-sa)

Species Name (Latin): spp.

Common Name: **Floribunda Rose**

Family Name (Latin and Common): *Rosaceae,* the Rose family

Related Family Members in Book: *Amelanchier, Chaenomeles, Malus, Prunus, Pyracantha, Pyrus,* and *Spiraea*

Availability: landscape plant: commercially limited as a cut flower

Color: pink

Unique Characteristics: Fragrant, multi-flowered clusters; different from typically available commercial varieties; many colors exist.

Design Applications: (small) (filler flower) Single-petal flowers are a dependable favorite flowering plant for the gardener.

Longevity: days 3-5 as a cut flower, with buds continuing to open

Genus Name (Latin): ***Rosa***   Pronunciation: (RO-sa)

Species Name (Latin): spp.

Common Name: **Rose**, Standard Rose, Tea Rose

Family Name (Latin and Common): *Rosaceae*, the Rose family

Related Family Members in Book: *Amelanchier, Chaenomeles, Malus, Prunus, Pyracantha, Pyrus,* and *Spiraea*

Availability: commercial product: year-round

Color: red, pink, yellow, white, lavender, orange, and bicolors

Unique Characteristics: Fragrant; air-dries well; will benefit from stems being cut under water during processing; flower stem has thorns; some varieties have ethylene sensitivity; many shades of colors exist; ecuadorian grown hybrids are typically large, both stem and flower head.

Design Applications: (medium to large) mass flower) The rose is as varied in color as in its applications. Uses are limitless. Long stem lengths of tea roses work well in parallel designs, with the flower form working in any style, always

adding mass. Roses are a common choice for bridal bouquets, corsages, and boutonnieres.

Longevity: days 3-10

Genus Name (Latin): ***Rosa***   Pronunciation: (RO-sa)

Species Name (Latin): spp.

Common Name: **Sweetheart Rose**

Family Name (Latin and Common): *Rosaceae*, the Rose family

Related Family Members in Book: *Amelanchier, Chaenomeles, Malus, Prunus, Pyracantha, Pyrus,* and *Spiraea*

Availability: commercial product: year-round

Color: red, white, yellow, orange, peach, and bicolors

Unique Characteristics: Fragrant; air-dries well; typically smaller flower head than standard roses; heavy water drinker; flower will benefit from stems being cut under water during processing; some varieties have ethylene sensitivity; many shades of colors exist.

Design Applications: (small–medium) (mass flower) Roses are an excellent choice for adding mass; it is also a good choice for bouquets, corsages, or boutonnieres. Like its larger counterpart, uses are limitless.

Longevity: days 3-6

Genus Name (Latin): ***Rosa***   Pronunciation: (RO-sa)

Species Name (Latin): spp.

Common Name: **Spray Rose**

Family Name (Latin and Common): *Rosaceae,* the Rose family

Related Family Members in Book: *Amelanchier, Chaenomeles, Malus, Prunus, Pyracantha, Pyrus,* and *Spiraea*

Availability: commercial product: year-round

Color: pinks, whites, and red

Unique Characteristics: Fragrant; air-dries well; stem is multi-flowered with buds in all stages of development; some varieties are ethylene sensitive.

Design Applications: (small) (filler flower) Multiflowered stem acts like a filler. Versatile, diminutive size is useful in feminine designs and garden-style mixed bouquets. Makes a nice, smaller choice for bouquets, corsages, and boutonnieres.

Longevity: days 4-7

Genus Name (Latin): ***Rudbeckia***   Pronunciation: (rud-BECK-i-a)

Species Name (Latin): *fulgida, hirta;* others and hybrids

Common Name: **Coneflower,** Black-Eyed Susan

Family Name (Latin and Common): *Asteraceae,* the Composite or Sunflower family

Related Family Members in Book: *Achillea, Ageratum, Artemisia, Aster, Bracteantha, Centaurea, Chrysanthemum, Cosmos, Craspedia, Dahlia, Echinacea, Echinops, Gerbera, Helianthus, Liatris, Senecio, Solidago,* x *Solidaster,* and *Zinnia*

Availability: commercial product: July-September

Color: orange, yellow

Unique Characteristics: Form is unique; cone-shaped black center is prominent.

Design Applications: (medium–large) (mass flower) Daisylike flower with contrasting center provides immediate accent, working well in combination with other summer bloomers.

Longevity: days 7-10

Genus Name (Latin): ***Saintpaulia***   Pronunciation: (saint-PAW-lee-a)

Species Name (Latin): spp.

Common Name: **African Violet**

Family Name (Latin and Common): *Gesneri-aceae,* the Gesneria family

Related Family Members in Book: *Sinningia*

Availability: indoor plant: year-round

Color: purple, violet, pink, red, yellow, blue, cream, white, bicolor, and multicolor

Unique Characteristics: Foliage is fleshy and can be feathered, flecked, or strongly varie-gated; flower petals may be ruffled; many cultivars exist.

Design Applications: (small) (accent) Delightful blooming houseplant is a nice addition of color and form to indoor container gardens.

Longevity: varies

Genus Name (Latin): ***Salix***   Pronunciation: (SAY-liks)

Species Name (Latin): ***discolor***

Common Name: **Pussy Willow**

Family Name (Latin and Common): *Salicaceae,* the Willow family

Related Family Members in Book: none

Availability: commercial product: varies, January-May

Color: white

Unique Characteristics: Commercial product has silky catkins, similar to *Salix* entry in dried material section of this book; photo shown is a mature specimen, available only from a landscape plant.

Design Applications: (small–large) (line flower) Branch material is useful in any design style, and is adaptable for armatures. Makes a nice addition to any spring flowered bouquet.

Longevity: varies by stage of catkin maturity

Genus Name (Latin): ***Salvia***   Pronunciation: (SAL-vee-a)

Species Name (Latin): ***farinacea* 'Victoria'**

Common Name: **Salvia,** Mealy-Cup Sage

Family Name (Latin and Common): *Lamiaceae,* the Mint family

Related Family Members in Book: *Lavandula, Leonotis, Mentha, Molucella, Monarda, Ocimum, Perovskia, Physostegia, Rosmarinus, Solenostemon, Stachys,* and *Thymus*

Availability: commercial product: varies July-September

Color: purple

Unique Characteristics: Air-dries well; contains strong color; inflorescence has slender spike-like form.

Design Applications: (medium) (filler flower) Linear filler provides spikes of intense color. Long, slender stem is a good choice for hand-tied bouquets.

Longevity: days 3-5 as a cut flower

Genus Name (Latin): ***Salvia***   Pronunciation: (SAL-vee-a)

Species Name (Latin): ***splendens* cultivar**

Common Name: **Salvia,** Sage

Family Name (Latin and Common): *Lamiaceae,* the Mint family

Related Family Members in Book: *Lavandula, Leonotis, Mentha, Molucella, Monarda, Ocimum, Perovskia, Physostegia, Rosmarinus, Solenostemon, Stachys,* and *Thymus*

Availability: landscape plant: varies commercially

Color: red (shown), deep purple, light salmon, lilac, and white

Unique Characteristics: Colors are bright on flower spikes with distinct calyxes; commercially available as annual; variety of cultivars available.

Design Applications: (medium) (color) Plant works well in container gardens for summer color, preferring full sun or partial shade. Works beautifully in combination with grey or complementary colored annuals.

Longevity: varies, as an annual

Genus Name (Latin): *Sandersonia*   Pronunciation: (san-der-SO-nee-a)

Species Name (Latin): *aurantiaca*

Common Name: *Sandersonia*

Family Name (Latin and Common): *Liliaceae,* the Lily family

Related Family Members in Book: *Asparagus, Aspidistra, Convallaria, Gloriosa, Hemerocallis, Hosta, Kniphotia, Liriopa, Muscari, Ornithogalum, Ruscus, Smilax, Tulipa,* and *Xerophyllum*

Availability: commercial product: varies July-May

Color: orange

Unique Characteristics: Form is distinctive; flower stem has pendent, urn-shaped florets; should be purchased with 4–5 full flowers and several immature buds; shows ethylene sensitivity.

Design Applications: (medium) (line flower) Unique flower form on slender stem adds color and accent. Use in contemporary styles

allowing plenty of room to show distinctive habit.

Longevity: days 10-14

Genus Name (Latin): *Saponaria*   Pronunciation: (sap-o-NAY-ree-a)

Species Name (Latin): *officinalis* cultivar

Common Name: *Saponaria,* Bouncing Bet, Soapwort

Family Name (Latin and Common): *Caryophyllaceae,* the Pink family

Related Family Members in Book: *Agrostemma, Dianthus,* and *Gypsophila*

Availability: commercial product: June-September

Color: pink, white

Unique Characteristics: Flowers are star-shaped at tips of branching stems.

Design Applications: (small) (filler flower) Delicate, airy appearance works well in mixed bouquets.

Longevity: days 7-10

Genus Name (Latin): ***Sarracenia***   Pronunciation: (sair-a-SEE-nee-a)

Species Name (Latin): *flava, purpurea;* cultivars

Common Name: ***Sarracenia,*** Pitcher Plant, Swamp Lily

Family Name (Latin and Common): *Sarraceniaceae,* the Pitcher Plant family

Related Family Members in Book: none

Availability: commercial product: April-September

Color: green

Unique Characteristics: Air-dries well (photo sample is dried); leaves are tubular and trumpet shaped, and are often mistaken for flowers; typically contains colored venation.

Design Applications: (medium) (line flower) This linear trumpet is wonderfully unique. Makes the perfect choice for parallel systems or simple lines in contemporary designs.

Longevity: days 7-10

Genus Name (Latin): ***Scabiosa***   Pronunciation: (skay-bi-OH-sa)

Species Name (Latin): *altropurpurea, caucasica;* cultivars

Common Name: ***Scabiosa,*** Pincushion Flower, Scabious

Family Name (Latin and Common): *Dipsacaceae,* the Teasel family

Related Family Members in Book: none

Availability: commercial product: June-October

Color: blue, white, yellow, or pink

Unique Characteristics: Flower has pincushion-like central florets with larger marginal ones; looking papery in appearance.

Design Applications: (small) (mass flower) A beautiful choice for adding mass to mixed arrangements. Works uniquely well in delicate or feminine-looking bouquets.

Longevity: days 6-10

Genus Name (Latin): *Schinus*   Pronunciation: (SKY-nus)

Species Name (Latin): *terebinthifolius, molle;*

Common Name: **Pepper Tree**, Brazilian Pepper Tree, Christmasberry Tree

Family Name (Latin and Common): *Anacardiaceae,* the Cashew family

Related Family Members in Book: *Cotinus*

Availability: commercial product: year-round

Color: red, pink

Unique Characteristics: Air-dries well; large berry cluster is pendulous along a leafy, almost weedy-appearing stem.

Design Applications: (small) (texture) Berry works well in combination with other natural materials for wreaths or with holiday foliages in decorative arrangements. Makes an excellent choice for adding textural interest to year-round designs.

Longevity: weeks 1-2

Genus Name (Latin): *Scilla*   Pronunciation: (SIL-a)

Species Name (Latin): spp.

Common Name: **Squill**

Family Name (Latin and Common): *Liliaceae,* the Lily family

Related Family Members in Book: *Asparagus, Aspidistra, Convallaria, Gloriosa, Hemerocallis, Hosta, Hyacinthus, Kniphotia, Liriopa, Muscari, Ornithogalum, Ruscus, Sandersonia, Smilax, Tulipa,* and *Xerophyllum*

Availability: commercial product: April-May

Color: lavender, blue, and white

Unique Characteristics: Form is similar to hyacinth.

Design Applications: (small) (filler flower) Bulb flower provides a unique form and texture to spring arrangements. Also useful for adding mass to any smaller scaled design.

Longevity: days 3-5

Genus Name (Latin): ***Sedum*** Pronunciation: (SEE-dum)

Species Name (Latin): **'Herbstfreude'** syn. *S.* 'Autumn Joy'

Common Name: ***Sedum,*** Stonecrop

Family Name (Latin and Common): *Crassulaceae,* the Orpine family

Related Family Members in Book: *Kalanchoe*

Availability: commercial product: May-October

Color: pink

Unique Characteristics: Color changes from pink during the summer months to an excellent burgundy color in fall.

Design Applications: (medium) (mass flower) Flower head provides visual mass to any design style, working well in autumnal arrangements once it has darkened to a bronze color.

Longevity: days 4-7

Genus Name (Latin): ***Serruria*** Pronunciation: (ser-ROO-ree-a)

Species Name (Latin): ***florida***

Common Name: ***Serruria,*** Blushing Bride

Family Name (Latin and Common): *Proteaceae,* the Protea family

Related Family Members in Book: *Banksia, Leucadendron, Leucospermum, Protea,* and *Telopea*

Availability: commercial product: varies June-July

Color: pink/salmon

Unique Characteristics: Form is unique with flower head having a cup-shaped ring of colored bracts.

Design Applications: (small) (filler flower) This flower is an excellent choice for texture. Unique appearance can be a nice filler when not overused.

Longevity: days 6-10

Genus Name (Latin): *Sinningia*   Pronunciation: (si-NIN-jee-a)

Species Name (Latin): spp.

Common Name: **Gloxinia**

Family Name (Latin and Common): *Gesneriaceae,* the Gesneria family

Related Family Members in Book: *Saintpaulia*

Availability: indoor plant: year-round

Color: red, purple, lavender, pink, white, and bicolors

Unique Characteristics: Foliage is fleshy, fragile; flowers have velvety appearance, and are tubular, trumpet shaped, or bell shaped; remove spent blossoms to allow light to hit emerging buds underneath; available as a greenhouse crop.

Design Applications: (large) (accent) Colorful blooming plant is beautiful. Makes a nice specimen to display alone for gifts or table décor.

Longevity: varies with buds continuing to open

Genus Name (Latin): *Solidago*   Pronunciation: (sol-i-DAY-go)

Species Name (Latin): spp.

Common Name: **Goldenrod,** Solidago

Family Name (Latin and Common): *Asteraceae,* the Composite or Sunflower family

Related Family Members in Book: *Achillea, Ageratum, Artemisia, Aster, Bracteantha, Centaurea, Chrysanthemum, Cosmos, Craspedia, Dahlia, Echinacea, Echinops, Gerbera, Helianthus, Liatris, Rudbeckia, Senecio, x Solidaster,* and *Zinnia*

Availability: commercial product: May-October

Color: yellow

Unique Characteristics: Air-dries well; often contains long enough laterals to be used individually.

Design Applications: (small) (filler flower) This yellow filler is good for everyday arrangements, sometimes containing a slender enough tip to be used as a line element in smaller-scaled designs.

Longevity: days 7-10

Genus Name (Latin): *x Solidaster*

Pronunciation: (so-li-DAS-ter)

Species Name (Latin): *luteus* syn. *S. hybridus*

Common Name: **Solidaster**

Family Name (Latin and Common): *Asteraceae,* the Composite or Sunflower family

Related Family Members in Book: *Achillea, Ageratum, Artemisia, Aster, Bracteantha, Centaurea, Chrysanthemum, Cosmos, Craspedia, Dahlia, Echinacea, Echinops, Gerbera, Helianthus, Liatris, Rudbeckia, Senecio, Solidago,* and *Zinnia*

Availability: commercial product: year-round

Color: yellow

Unique Characteristics: Air-dries well; has daisylike flower heads.

Design Applications: (small) (filler flower) Tiny flowered filler is a good choice for everyday designs. Provides nice addition to autumnal bouquets. Useful accent for corsages or boutonnieres.

Longevity: days 7-10

Genus Name (Latin): *Spiraea*

Pronunciation: (spy-REE-a)

Species Name (Latin): spp.

Common Name: **Spirea**

Family Name (Latin and Common): *Rosaceae,* the Rose family

Related Family Members in Book: *Amelanchier, Chaenomeles, Malus, Prunus, Pyracantha, Pyrus,* and *Rosa*

Availability: landscape plant: commercially limited as a cut flower.

Color: white

Unique Characteristics: Stem has arching habit; displays a delicate appearance.

Design Applications: (small) (filler flower) Clusters of tiny flowers, along arching stem, can provide filler to line-mass designs.

Longevity: varies

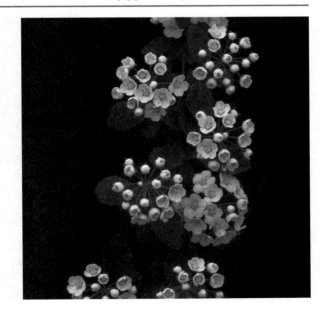

Genus Name (Latin): *Stachys*

Pronunciation: (STAK-is)

Species Name (Latin): *byzantina*

Common Name: **Lamb's Ear,** Woolly Betony

Family Name (Latin and Common): *Lamiaceae,* the Mint family

Related Family Members in Book: *Lavandula, Leonotis, Melissa, Mentha, Molucella, Monarda, Ocimum, Perovskia, Physostegia, Rosmarinus, Solenostemon,* and *Thymus*

Availability: commercial product: varies May-July

Color: lavender/pink

Unique Characteristics: Foliage is woolly and gray-green in color; leaves are a good choice for pressing.

Design Applications: (medium) (line flower) Great line material to provide interesting texture and form.

Longevity: varies

Genus Name (Latin): *Stephanotis*

Pronunciation: (stef-a-NO-tis)

Species Name (Latin): *floribunda*

Common Name: **Stephanotis,** Madagascar Jasmine

Family Name (Latin and Common): *Asclepi-adaceae,* the Milkweed family

Related Family Members in Book: *Asclepias, Tweedia*

Availability: commercial product: year-round

Color: white

Unique Characteristics: Fragrance is sweet; requires special wiring techniques to be used in design work; stem is slender and tubular approximately 1" to 1 1/2" in length.

Design Applications: (small) (form flower) Small, star-shaped flower is used primarily for wedding designs; Especially useful for corsages, boutonnieres, and headpieces. It can look beautiful in mass, also adding to the concentration of its sweet fragrance.

Longevity: days 3-4

Genus Name (Latin): ***Strelitzia***   Pronunciation: (stre-LIT-see-a)

Species Name (Latin): *reginae*

Common Name: **Bird-of-Paradise**, Crane Flower

Family Name (Latin and Common): *Strelitzi-aceae,* the Strelitzia family

Related Family Members in Book: none

Availability: commercial product: year-round

Color: orange and blue

Unique Characteristics: Flower is distinctive resembling a bird; stem is thick and leafless; remove old florets from spathe and continue to lift out new ones.

Design Applications: (large) (form flower) Dramatic flower commands immediate attention. Use in contemporary designs in combination with other exotics or tropicals. Excellent choice for geometric styles. Allow plenty of room to display unique character.

Longevity: weeks 1-2

Genus Name (Latin): ***Symphoricarpos***   Pronunciation: (sim-fo-ri-KAR-pus)

Species Name (Latin): ***albus* var. *laevigatus***

Common Name: **Snowberry**

Family Name (Latin and Common): *Caprifoli-aceae,* the Honeysuckle family

Related Family Members in Book: *Kolkwitzia, Lonicera,* and *Viburnum*

Availability: commercial product: varies November-December

Color: white

Unique Characteristics: Fruit is spherical and pure white.

Design Applications: (medium) (texture) Berry is an excellent texture. Use in traditional and contemporary design styles.

Longevity: weeks 1-2

Genus Name (Latin): *Syringa* — Pronunciation: (si-RING-ga)

Species Name (Latin): *vulgaris;* others and cultivars

Common Name: **Lilac,** Common Lilac

Family Name (Latin and Common): *Oleaceae,* the Olive family

Related Family Members in Book: *Forsythia*

Availability: commercial product: January-May

Color: white, lilac, pink, and purple

Unique Characteristics: Fragrance is sweet; showy panicles of tiny star-shaped florets; also is a heavy water drinker.

Design Applications: (large) (mass flower) Visually heavy panicles of fragrant florets are nice additions to spring bouquets and are beautiful when used in mass.

Longevity: days 3-10

Genus Name (Latin): *Tagetes* — Pronunciation: (ta-JEE-tee-z)

Species Name (Latin): *erecta, patula;* others and hybrids

Common Name: **Marigold**

Family Name (Latin and Common): *Asteraceae,* the Composite or Sunflower family

Related Family Members in Book: *Achillea, Ageratum, Artemisia, Aster, Bracteantha, Centaurea, Chrysanthemum, Cosmos, Craspedia, Dahlia, Echinacea, Echinops, Gerbera, Helianthus, Liatris, Rudbeckia, Senecio, Solidago, x Solidaster,* and *Zinnia*

Availability: commercial product: July-September

Color: orange, yellow, and bicolors

Unique Characteristics: Fragrant; some people find scent offensive.

Design Applications: (medium) (mass flower) The marigold provides mass with its densely petaled form. Makes a colorful companion for other bright summer bloomers.

Longevity: days 4-6

Genus Name (Latin): ***Tanacetum***   Pronunciation: (tan-ah-SEE-tum)

Species Name (Latin): spp.

Common Name: **Feverfew**

Family Name (Latin and Common): *Asteraceae,* the Composite or Sunflower family

Related Family Members in Book: *Achillea, Ageratum, Artemisia, Aster, Bracteantha, Centaurea, Cosmos, Craspedia, Dahlia, Echinacea, Echinops, Gerbera, Helianthus, Liatris, Rudbeckia, Senecio, Solidago, x Solidaster,* and *Zinnia*

Availability: commercial product: year-round

Color: white

Unique Characteristics: Fragrant; contact with skin may aggravate skin allergies.

Design Applications: (small) (filler flower) Dainty looking filler is useful in feminine designs.

Longevity: days 7-10

Genus Name (Latin): ***Telopea***   Pronunciation: (te-LO-pee-ah)

Species Name (Latin): ***speciosissima***

Common Name: ***Telopea,*** Waratah

Family Name (Latin and Common): *Proteaceae,* the Protea family

Related Family Members in Book: *Banksia, Leucadendron, Leucospermum,* and *Serruria*

Availability: commercial product: May-November

Color: red

Unique Characteristics: Flower and foliage have exotic appearance; terminal racemes are surrounded by colorful bracts.

Design Applications: (large) (form flower) Flower should be placed in focal zones to highlight unique form, best if used in combination with other exotics.

Longevity: weeks 1-2

Genus Name (Latin): *Thunbergia*

Pronunciation: (thun-BUR-jee-a)

Species Name (Latin): *alata*

Common Name: **Black-Eyed Susan Vine,** Clock Vine

Family Name (Latin and Common): *Acanthaceae,* the Acanthus family

Related Family Members in Book: *Pachysandra*

Availability: landscape plant: varies

Color: orange

Unique Characteristics: Plant is available as an annual and has a vining habit.

Design Applications: (small) (accent) Flowering vine is a nice choice for adding accent to hanging baskets or mixed container gardens. Can grow to cover arbors or decorations of 5–6 feet high, making it useful for several types of display.

Longevity: varies as a bedding plant

Genus Name (Latin): *Trachelium*

Pronunciation: (tra-KEE-lee-um)

Species Name (Latin): *caeruleum*

Common Name: *Trachelium,* Blue Throatwort, Throatwort

Family Name (Latin and Common): *Campanulaceae,* the Bellflower family

Related Family Members in Book: *Campanula*

Availability: commercial product: March-November

Color: purple

Unique Characteristics: Flower head consists of dense, terminal corymbose clusters.

Design Applications: (medium–large) (mass flower) Useful flower for traditional line-mass designs. Round flower head is a good candidate for the pillowing technique.

Longevity: days 6-10

Genus Name (Latin): ***Trachymene*** syn. *Didiscus*   Pronunciation: (tra-KYM-e-nee)

Species Name (Latin): *coerulea*

Common Name: ***Didiscus,*** Blue Laceflower, Lace-flower

Family Name (Latin and Common): *Apiaceae,* the Parsley or Carrot family

Related Family Members in Book: *Ammi*

Availability: commercial product: June-November

Color: blue

Unique Characteristics: Flower has delicate appearance; buds are similar in appearance to Queen Ann's Lace; flower head is similar to *Scabiosa.*

Design Applications: (small) (filler flower) Round flower has delicate appearance; works well in romantic styles, and is a nice choice for filler in mixed bouquets.

Longevity: days 7-10

Genus Name (Latin): ***Tradescantia***   Pronunciation: (trad-es-KAN-shee-a)

Species Name (Latin): **x Andersoniana Group**

Common Name: **Spiderwort**

Family Name (Latin and Common): *Commelinaceae,* the Spiderwort family

Related Family Members in Book: none

Availability: landscape plant: commercially limited as a cut flower.

Color: purple

Unique Characteristics: Flower borne in paired, terminal cymes; has a "spidery" appearance; florets are short-lived with buds continuing to open; prefers clear water to floral foam.

Design Applications: (medium) (form flower) Flower head works well in abstract designs where unique form can be highlighted.

Longevity: varies, several days per bud

Genus Name (Latin): *Triteleia*   Pronunciation: (tri-te-LAY-a)

Species Name (Latin): sp.

Common Name: *Triteleia,* Brodiaea

Family Name (Latin and Common): *Liliaceae,* the Lily family

Related Family Members in Book: *Asparagus, Aspidistra, Convallaria, Gloriosa, Hemerocallis, Hosta, Hyacinthus, Kniphofia, Liriope, Muscari, Ruscus, Sandersonia, Scilla, Smilax, Tulipa,* and *Xerophyllum*

Availability: commercial product: May-November

Color: violet, white, and purple-blue

Unique Characteristics: Form is similar in appearance to *Agapanthus.*

Design Applications: (small) (mass flower) This flower is a useful accent in smaller-scaled arrangements.

Longevity: days 7-10

Genus Name (Latin): *Triticum*   Pronunciation: (TRIT-i-kum)

Species Name (Latin): spp.

Common Name: **Wheat**

Family Name (Latin and Common): *Poaceae,* the Grass family

Related Family Members in Book: *Miscanthus, Pennisetum, Phalaris, Setaria,* and *Sorghum*

Availability: commercial product: May-October

Color: green

Unique Characteristics: Seed head provides interesting texture, color, and form.

Design Applications: (small) (line flower) Wheat is most effective when grouped in mass. Excellent in combination with vegetables and fruit. When bundled, the stems work very well in autumn harvest designs or monochromatic arrangements of any style. Makes a good candidate for parallel systems.

Longevity: days 5-7

Genus Name (Latin): *Tulipa*   Pronunciation: (TYOU-li-pa)

Species Name (Latin): spp.

Common Name: **Tulip**

Family Name (Latin and Common): *Liliaceae,* the Lily family

Related Family Members in Book: *Asparagus, Aspidistra, Convallaria, Gloriosa, Hosta, Hyacinthus, Kniphofia, Liriope, Muscari, Ruscus, Sandersonia, Scilla, Smilax,* and *Xerophyllum*

Availability: commercial product: November-May

Color: red, yellow, purple, pink, orange, peach, white, and bicolors

Unique Characteristics: Stem continues to lengthen or "grow" even after being placed into designs; flowers are available as field grown or French varieties (French tulips typically have longer flower stem and flower head); many hybrids and cultivars are available, some with fringed or striped sepals or petals.

Design Applications: (medium) (mass flower) Rounded flower always provides interesting mass. Like most bulbs, the tulip works best in combination with other spring flowers, but is

exceptionally beautiful in mass on their own. Use in vegetative systems; popular in white and red during the winter holiday season.

Longevity: days 3-7

Genus Name (Latin): *Tweedia* syn. *Oxypetalum*   Pronunciation: (TWEE-dee-a)

Species Name (Latin): *caerulea*

Common Name: *Tweedia*

Family Name (Latin and Common): *Asclepiadaceae,* the Milkweed family

Related Family Members in Book: *Stephanotis*

Availability: commercial product: limited

Color: blue

Unique Characteristics: Flower displays hairy, white stems; milky sap is exuded when cut; flower has distinctive sky blue color.

Design Applications: (small) (accent) Flowers of intense blue color add accent and are a nice combination with pink materials.

Longevity: days 3-5

Genus Name (Latin): *Vanda*   **Pronunciation:** (VAN-da)

Species Name (Latin): *caerulea, tessellata, tricolor;* others and many hybrids

Common Name: *Vanda*

Family Name (Latin and Common): *Orchidaceae,* the Orchid family

Related Family Members in Book: *Arachnis, Cattleya, Cymbidium, Dendrobium, Oncidium, Paphiopedilum,* and *Phalaenopsis*

Availability: commercial product: varies year-round

Color: purple, burgundy, red-brown, dark violet, mauve, many other combinations

Unique Characteristics: Flower can be fragrant; wide assortment of rich colors including blue rarely found in the *Orchidaceae* family; some contain heavily spotted patterns.

Design Applications: (medium) (mass flower) Sprays of several florets are dramatic and best if allowed plenty of room to display their unique form. Makes an excellent candidate for

high-style designs. Individual blossoms well suited for corsages or boutonnieres.

Longevity: weeks 1-3

Genus Name (Latin): *Verbena* syn. *Glandularia*   **Pronunciation:** (vur-BEE-na)

Species Name (Latin): *bonariensis*

Common Name: *Verbena*

Family Name (Latin and Common): *Verbenaceae,* the Vervain or Verbena family

Related Family Members in Book: *Callicarpa, Lantana*

Availability: commercial product: limited

Color: lavender-purple

Unique Characteristics: Stem is slender and leafless with salverform flowers in panicle-like cymes.

Design Applications: (small) (filler flower) Flower head on slender stem allows this *Verbena* to be used successfully in parallel designs, while acting as a filler in other styles. Use for adding mass to small arrangements.

Longevity: days 4-7 as a cut flower

Genus Name (Latin): ***Verbena*** syn. *Glandularia*   Pronunciation: (vur-BEE-na)

Species Name (Latin): ***x hybrida***

Common Name: ***Verbena***

Family Name (Latin and Common): *Verbenaceae,* the Vervain or Verbena family

Related Family Members in Book: *Callicarpa, Lantana*

Availability: landscape plant: varies commercially

Color: white, pinks, purple, red, peach, lavender, yellow, and blue

Unique Characteristics: Colors are many; typically sold as an annual bedding plant.

Design Applications: (small) (accent) Clusters of multiple, colorful florets produce mass. Add to container gardens where the often trailing habit of this plant is visible.

Longevity: varies

Genus Name (Latin): ***Veronica***   Pronunciation: (ve-RON-i-ka)

Species Name (Latin): ***spicata*** cultivar

Common Name: ***Veronica,*** Spike Speedwell

Family Name (Latin and Common): *Scrophulariaceae,* the Figwort family

Related Family Members in Book: *Antirrhinum, Digitalis, Leucophyllum,* and *Penstemon*

Availability: commercial product: year-round, peak supplies June–September

Color: rose, white, purple, and blue

Unique Characteristics: Flower tips gently curve; displays spikelike racemes of tiny flowers and toothed leaves.

Design Applications: (medium) (filler flower) This linear filler is a good choice for summer garden bouquets.

Longevity: days 5-7

Genus Name (Latin): ***Viburnum***   Pronunciation: (vy-BUR-num)

Species Name (Latin): ***carlesii***

Common Name: **Koreanspice Viburnum**

Family Name (Latin and Common): *Caprifoli-aceae,* the Honeysuckle family

Related Family Members in Book: *Kolkwitzia, Lonicera, Symphoricarpus,* and *Weigela*

Availability: landscape plant: not available commercially as a cut flower.

Color: white

Unique Characteristics: Fragrance is strong.

Design Applications: (medium) (mass flower) Flower head can stand on its own, providing wonderful fragrance. Nice choice for Ikebana.

Longevity: days 2-5

Genus Name (Latin): ***Viburnum***   Pronunciation: (vy-BUR-num)

Species Name (Latin): ***davidii***

Common Name: **Viburnum**

Family Name (Latin and Common): *Caprifoli-aceae,* the Honeysuckle family

Related Family Members in Book: *Kolkwitzia, Lonicera, Symphoricarpus,* and *Weigela*

Availability: commercial product: July–December

Color: blue (metallic)

Unique Characteristics: Color is unique.

Design Applications: (small) (filler) Choice berry is beautiful, and an excellent material for adding texture and accent.

Longevity: weeks 1-2 or longer

Genus Name (Latin): *Viburnum* Pronunciation: (vy-BUR-num)

Species Name (Latin): *opulus*

Common Name: **European Cranberrybush Viburnum**

Family Name (Latin and Common): *Caprifoliaceae,* the Honeysuckle family

Related Family Members in Book: *Kolkwitzia, Lonicera, Symphoricarpus* and *Weigela*

Availability: commercial product: varies May–June

Color: white

Unique Characteristics: Color is spring green before turning white; varying sized flower clusters on stem; leaf form is unique; flower is a heavy water drinker.

Design Applications: (small–medium) (mass flower) Blossoms are excellent specimens to be used in mass. Good choice for providing visual weight to garden-style designs.

Longevity: days 5-7

Genus Name (Latin): *Viburnum* Pronunciation: (vy-BUR-num)

Species Name (Latin): *x rhytidophylloides*

Common Name: *Lantanaphyllum Viburnum*

Family Name (Latin and Common): *Caprifoliaceae,* the Honeysuckle family

Related Family Members in Book: *Kolkwitzia, Lonicera, Symphoricarpus,* and *Weigela*

Color: white

Availability: landscape plant: typically not available as a commercial cut flower

Unique Characteristics: Flower has a fragrance that some find offensive; large leathery leaf is extremely tomentose beneath; typically is available from a spring flowering landscape plant.

Design Applications: (large) (mass flower) Flower is useful in Ikebana designs.

Longevity: varies

150 . . . . *Flowers and Fruit*

Genus Name (Latin): *Viola*   Pronunciation: (VY-oh-la)

Species Name (Latin): ***tricolor*** cultivar

Common Name: **Johnny-jump-up**

Family Name (Latin and Common): *Violaceae*, the Violet family

Related Family Members in Book: *Viola*

Availability: landscape plant: varies commercially March-April

Color: purple, violet, yellow, white, and bicolors

Unique Characteristics: Plant is typically available as an annual; a favorite of Elizabethan times, this flower can be pressed; used often in rock gardens or in combination with other early flowering bulb plantings.

Design Applications: (small) (accent) Dainty flower is a delightful specimen for the tussie-mussie. Makes a great choice for pressed floral designs; plant is a nice addition to containerized spring gardens.

Longevity: days 2-4 as a cut flower; preferring clear water to floral foam.

Genus Name (Latin): *Viola*   Pronunciation: (VY-oh-la)

Species Name (Latin): **x *Wittrockiana*** cultivars

Common Name: **Pansy**

Family Name (Latin and Common): *Violaceae*, the Violet family

Related Family Members in Book: *Viola*

Availability: landscape plant: varies commercially March-April

Color: purple, yellow, lavender, white, orange, burgundy, and bicolors

Unique Characteristics: Flower appears to have a delightful "face"; loving cooler temperatures, this annual is one of the first available in spring; excellent candidate for pressing.

Design Applications: (small) (accent) This exceptional flower will make you smile. Complements spring containerized gardens. Nice choice for early spring indoor displays.

Longevity: days 3-5 as a cut flower; preferring clear water to floral foam.

Genus Name (Latin): ***Watsonia***   Pronunciation: (wot-SO-ni-a)

Species Name (Latin): sp.

Common Name: ***Watsonia,*** Bugle Lily

Family Name (Latin and Common): *Iridaceae,* the Iris family

Related Family Members in Book: *Crocosmia, Gladiolus, Iris,* and *Ixia*

Availability: commercial product: April-June

Color: white

Unique Characteristics: Flower is linear and similar to traditional *Gladiolus.*

Design Applications: (large) (line flower) This material works well in line mass designs. Exceptionally nice when used alone in massed bundles.

Longevity: days 5-7

Genus Name (Latin): ***Weigela***   Pronunciation: (wy-GEE-la)

Species Name (Latin): spp.

Common Name: ***Weigela***

Family Name (Latin and Common): *Caprifoliaceae,* the Honeysuckle family

Related Family Members in Book: *Kolkwitzia, Lonicera, Symphoricarpus,* and *Viburnum*

Availability: landscape plant: commercially available May-June as a cut flower

Color: pink, purplish, carmine, and white

Unique Characteristics: Stem is woody; flowers appear on prior year's growth of plant.

Design Applications: (medium) (line flower) Leafy line material is nice in combination with other late spring bloomers.

Longevity: days 3-5 as a cut flower

Genus Name (Latin): ***Wisteria*** Pronunciation: (wis-TEE-ri-a; wis-TEER-i-a)

Species Name (Latin): sp.

Common Name: ***Wisteria***

Family Name (Latin and Common): *Fabaceae,* the Pea or Pulse family

Related Family Members in Book: *Acacia, Cercis, Cytsus, Lathyrus,* and *Lupinus*

Availability: landscape plant: commercially limited as a cut flower

Color: violet, violet-blue, white, purple, and reddish-violet

Unique Characteristics: Flower is fragrant, pendent, and of mostly terminal racemes; typically available from a spring flowering vine.

Design Applications: (large) (accent) This beautiful flower is best if pendulous character is highly visible, for example, along an arbor or constructed armature.

Longevity: varies

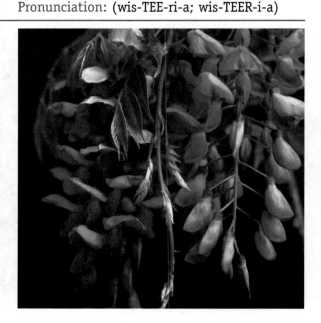

Genus Name (Latin): ***Xeranthemum*** Pronunciation: (zee-RAN-the-mum)

Species Name (Latin): ***annuum***

Common Name: ***Xeranthemum,*** Immortelle

Family Name (Latin and Common): *Asteraceae,* the Composite or Sunflower family

Related Family Members in Book: *Achillea, Ageratum, Artemisia, Aster, Bracteantha, Centaurea, Chrysanthemum, Cosmos, Craspedia, Dahlia, Echinacea, Echinops, Gerbera, Helianthus, Liatris, Rudbeckia, Senecio, Solidago,* x *Solidaster,* and *Zinnia*

Availability: commercial product: July-October

Color: pink, white

Unique Characteristics: Air-dries well; petals have papery appearance.

Design Applications: (small) (filler flower) Delicate filler is beautiful in mixed bouquets.

Longevity: days 7-10

Genus Name (Latin): *Yucca*   Pronunciation: (YUK-a)

Species Name (Latin): spp.

Common Name: *Yucca*

Family Name (Latin and Common): *Agavaceae,* the Agave family

Related Family Members in Book: *Cordyline, Dracaena, Polianthes,* and *Phormium*

Availability: landscape plant: commercially limited June-July as a cut flower

Color: white

Unique Characteristics: Stem of flower is thick and leafless, with florets that continue to open.

Design Applications: (large) (line material) Use for accent in large designs or displays.

Longevity: days 5-7 as a cut flower

Genus Name (Latin): *Zantedeschia*   Pronunciation: (zan-tee-DES-ki-a)

Species Name (Latin): *aethiopica*

Common Name: **Calla,** Calla Lily

Family Name (Latin and Common): *Araceae,* the Arum family

Related Family Members in Book: *Anthurium, Caladium, Monstera,* and *Philodendron*

Availability: commercial product: year-round

Color: white

Unique Characteristics: Flower is not a true "lily" but is often referred to as such; spathe of color surrounds the spadix; 'Green Goddess' cultivar of this species is exceptional.

Design Applications: (large) (line material) The graceful 'calla' is a classic. Use often in wedding and bridal designs. It is beautiful in mass.

Longevity: days 4-10

Genus Name (Latin): ***Zantedeschia***   Pronunciation: (zan-tee-DES-ki-a)

Species Name (Latin): ***elliotiana***

Common Name: **Calla,** Golden Calla, Yellow Calla

Family Name (Latin and Common): *Araceae,* the Arum family

Related Family Members in Book: *Anthurium, Caladium, Monstera,* and *Philodendron*

Availability: commercial product: year-round

Color: yellow

Unique Characteristics: Flower is not a true "lily," but is often referred to as such; spathe of color surrounds the spadix.

Design Applications: (medium–large) (line material) The 'calla' is a classic. Stems can be gently bent for graceful curved lines in contemporary and traditional design styles.

Longevity: days 4-10

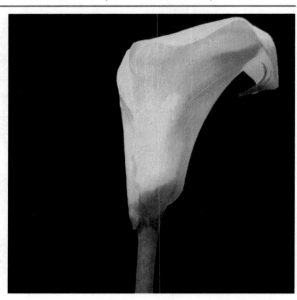

Genus Name (Latin): ***Zantedeschia***   Pronunciation: (zan-tee-DES-ki-a)

Species Name (Latin): ***rehmannii***

Common Name: **Calla,** Pink Calla

Family Name (Latin and Common): *Araceae,* the Arum family

Related Family Members in Book: *Anthurium, Caladium, Monstera,* and *Philodendron*

Availability: commercial product: year-round

Color: pink

Unique Characteristics: Flower is not a true "lily," but is often referred to as such; spathe of color surrounds the spadix; this species is often smaller than *Z. elliotiana* and *Z. aethiopica.*

Design Applications: (medium) (line flower) The classic line of the 'calla' is often used in bridal designs. Wonderful line material. Can easily be shaped to create curving lines in contemporary design techniques.

Longevity: days 4-10

Genus Name (Latin): ***Zingiber***   Pronunciation: (ZIN-ji-bur)

Species Name (Latin): sp.

Common Name: **Shampoo Ginger**

Family Name (Latin and Common): *Zingiber-aceae,* the Ginger family

Related Family Members in Book: *Alpinia, Costus, Curcuma, Globba*

Availability: commercial product: June-September

Color: yellow

Unique Characteristics: Flower appears inside waxy bracts; has a pinecone shape.

Design Applications: (medium) (form flower) Unique flower provides immediate accent with unusual shape and combines well with other exotic or tropical flowers.

Longevity: days 7-10

Genus Name (Latin): ***Zinnia***   Pronunciation: (ZIN-ee-a)

Species Name (Latin): *elegans* cultivar

Common Name: *Zinnia*

Family Name (Latin and Common): *Asteraceae,* the Composite or Sunflower family

Related Family Members in Book: *Achillea, Ageratum, Artemisia, Aster, Bracteantha, Centaurea, Chrysanthemum, Cosmos, Craspedia, Dahlia, Echinacea, Echinops, Gerbera, Helianthus, Liatris, Rudbeckia, Senecio, Solidago,* and *x Solidaster*

Availability: commercial product: June-September

Color: orange, pink, yellow, red, and white

Unique Characteristics: Colors are bold.

Design Applications: (medium) (mass flower) Round flower adds accent and mass to traditional or contemporary designs. Combines well with other bold summer bloomers.

Longevity: days 5-7

Genus Name (Latin): ***Abelmoschus***

Pronunciation: (ab-el-MOS-kus)

Species Name (Latin): *esculentus*

Common Name: **Okra,** Lady's-Finger

Family Name (Latin and Common): *Malvaceae,* the Mallow family

Related Family Members in Book: *Hibiscus, Lavatera*

Availability: commercial product: year-round

Color: tan with brown striations

Unique Characteristics: Dried material; Fruit of okra is 5-angled-cylindrical, beaked and bristly; is commercially available dried and picked on sticks.

Design Applications: (medium) (form) Finger-shaped form adds accent and texture to designs. Nice addition to autumnal center-pieces.

Longevity: long lasting

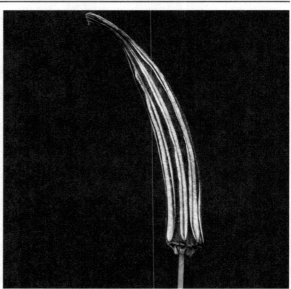

Genus Name (Latin): ***Abies***

Pronunciation: (AY-beez)

Species Name (Latin): *alba*

Common Name: **Silver Fir**

Family Name (Latin and Common): *Pinaceae,* the Pine family

Related Family Members in Book: *Larix, Picea, Pinus, Pseudotsuga,* and *Tsuga*

Availability: commercial product: October-January

Color: green (dark) with underside of needles being silver

Unique Characteristics: Needle is a distinctive silver on underside.

Design Applications: (small) (texture) Short needled evergreen is an excellent filler. Beautiful in combination with other mixed conifers.

Longevity: weeks 3-4 long lasting

Genus Name (Latin): ***Abies***   Pronunciation: (AY-beez)

Species Name (Latin): ***balsamea***

Common Name: **Balsam Fir**

Family Name (Latin and Common): *Pinaceae,*
the Pine family

Related Family Members in Book: *Larix, Picea,*
Pinus, Pseudotsuga, and *Tsuga*

Availability: commercial product: October-
January

Color: green (dark)

Unique Characteristics: Needle is short.

Design Applications: (small) (texture) Short
branched tips provide good texture in combination with other conifers.

Longevity: weeks 3-4 long lasting

Genus Name (Latin): ***Abies***   Pronunciation: (AY-beez)

Species Name (Latin): ***procera***

Common Name: **Noble Fir**

Family Name (Latin and Common): *Pinaceae,*
the Pine family

Related Family Members in Book: *Larix, Picea,*
Pinus, Pseudotsuga, and *Tsuga*

Availability: commercial product: October-
January

Color: green (bluish green)

Unique Characteristics: Needles are stiff and
dense along rigid stems; also available
preserved with glycerine.

Design Applications: (small) (texture) Short
needled evergreen provides sturdy basis for
holiday centerpieces and decorations. Works
beautifully in combination with other conifers.

Longevity: weeks 3-4 long lasting

Genus Name (Latin): ***Acacia***   Pronunciation: (a-KAY-sha)

Species Name (Latin): ***cultriformis***

Common Name: **Knifeblade Acacia,** Knife Acacia

Family Name (Latin and Common): *Fabaceae,* the Pea or Pulse family

Related Family Members in Book: *Cercis, Cytisus, Lathyrus, Lupinus,* and *Wisteria*

Availability: commercial product: varies year-round

Color: gray-green

Unique Characteristics: Shape of leaf is knifelike; often contains clustered flower heads within long racemes at end of bracts.

Design Applications: (small) (texture) Leafed stem is distinctive. Different portions of stem can be used in various ways, adding texture, accent, or line.

Longevity: days 7-10

Genus Name (Latin): ***Acacia***   Pronunciation: (a-KAY-sha)

Species Name (Latin): ***retinoides***

Common Name: **Acacia,** Silver Wattle, Swamp Wattle, Wirilda

Family Name (Latin and Common): *Fabaceae,* the Pea or Pulse family

Related Family Members in Book: *Cercis, Cytisus, Lathyrus, Lophomyrtus,* and *Wisteria*

Availability: commercial product: varies October-January

Color: green (bluish-green)

Unique Characteristics: Leaf is distinctly narrow and lance shaped; lemon yellow flower heads are bold and often present, but leaves typically dominate overall appearance.

Design Applications: (small) (filler) Skinny leaf is an excellent choice for texture and filler. Makes a nice addition for monochromatic color schemes. Individual leaves are useful in body flower designs.

Longevity: weeks 1-2

Genus Name (Latin): ***Acalypha***   Pronunciation: (ak-a-LEE-fa; ak-a-LY-fa)

Species Name (Latin): ***wilkesiana*** cultivar

Common Name: **Copperleaf,** Jacob's Coat, Fire-Dragon, Beefsteak Plant

Family Name (Latin and Common): *Euphorbiaceae,* the Spurge family

Related Family Members in Book: *Codiaeum, Euphorbia,* and *Sapium*

Availability: indoor plant: commercially limited as a cut foliage

Color: yellow/green (shown)

Unique Characteristics: Colors are widely varied and mottled; plant is mostly known for its unique coloration and typically is grown as a tropical foliage plant; several cultivars available.

Design Applications: (large) (accent) Leaf with varied colors is an excellent accent to containerized plantings or temporary displays. A good specimen for adding mass.

Longevity: varies as a cut foliage

Genus Name (Latin): ***Acer***   Pronunciation: (AY-sur)

Species Name (Latin): ***palmatum*** cultivar

Common Name: **Japanese Maple**

Family Name (Latin and Common): *Aceraceae,* the Maple family

Related Family Members in Book: none

Availability: landscape plant: commercial availability limited May-November as a cut foliage

Color: greens, reds, reddish-purple, and yellowish-orange

Unique Characteristics: Foliage is deeply 5-7 lobed with serrate margins; wonderful leaf to press; periodically appears commercially with colors varying by season and cultivar.

Design Applications: (medium) (texture) Unusual-shape leaf, provides excellent texture and accent. Makes a nice choice for Ikebana designs.

Longevity: varies

Genus Name (Latin): ***Actinidia***   Pronunciation: (ak-ti-NID-ee-a)

Species Name (Latin): ***deliciosa***

Common Name: **Kiwi Fruit Vine**

Family Name (Latin and Common): *Actinidiaceae,* the Actinidia family

Related Family Members in Book: none

Availability: commercial product: year-round

Color: brown (light)

Unique Characteristics: Dried material; stem of fruit has gnarled appearance; *Actinidia* is cultivated as an ornamental vine for its edible fruit.

Design Applications: (large) (line) Wonderfully curving vine works well in contemporary perishable or dried container arrangements.

Longevity: long lasting

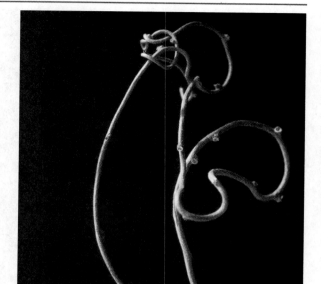

Genus Name (Latin): ***Adiantum***   Pronunciation: (ad-i-AN-tum)

Species Name (Latin): spp.

Common Name: **Maidenhair Fern,** Maidenhair

Family Name (Latin and Common): *Polypodiaceae,* the Polypody family

Related Family Members in Book: *Nephrolepsis, Platycerium, Polystichum,* and *Rumohra*

Availability: commercial product: year-round

Color: green (light)

Unique Characteristics: Stem is typically black; holds fan-shaped pinnules; also available preserved with glycerine.

Design Applications: (small) (form) Delicate-looking fern is an excellent choice for feminine designs, and a good foliage for providing interest.

Longevity: days 3-5

Genus Name (Latin): **Aegopodium**   Pronunciation: (ee-go-PO-dee-um)

Species Name (Latin): **podograria 'Variegatum'**

Common Name: **Goutweed,** Bishop's Weed, Ground Ash, Ground Elder

Family Name (Latin and Common): *Apiaceae,* the Parsley or Carrot family

Related Family Members in Book: *Ammi, Anethum, Bupleurum, Eryngium,* and *Trachymene*

Availability: landscape plant: typically not available as a commercial cut foliage

Color: green (white margins)

Unique Characteristics: Foliage is available from a perennial used for edging and ground cover.

Design Applications: (small) (filler) Variegated filler brightens focal zones as an addition to foliage bouquets.

Longevity: days 2-4

Genus Name (Latin): **Ajania**   Pronunciation: (a-JAY-nee-a)

Species Name (Latin): **pacifica** syn. *Chrysanthemum pacificum, Dendranthema pacificum*

Common Name: **Ajania**

Family Name (Latin and Common): *Asteraceae,* the Composite or Sunflower family

Related Family Members in Book: *Achillea, Ageratum, Artemisia, Aster, Bracteantha, Centaurea, Chrysanthemum, Cosmos, Craspedia, Dahlia, Echinacea, Echinops, Gerbera, Helianthus, Leucanthemum, Liatris, Rudbeckia, Senecio, x Solidaster,* and *Zinnia*

Availability: indoor plant: varies commercially

Color: green with white margins

Unique Characteristics: Foliage is unique and distinctive; In autumn, small yellow flowers appear in branched corymbs.

Design Applications: (small) (accent) Unusual plant is an excellent choice for accent in temporary and permanent landscape displays. Makes a useful filler for cut-flower bouquets.

Works especially well in designs of complete foliage.

Longevity: varies

Genus Name (Latin): *Alpinia*

Pronunciation: (al-PIN-ee-a)

Species Name (Latin): *zerumbet* 'Variegata'

Common Name: **Variegated Ginger**, Variegated Ginger Lily

Family Name (Latin and Common): *Zingiberaceae,* the Ginger family

Related Family Members in Book: *Costus, Curcuma,* and *Globba*

Availability: commercial product: year-round

Color: yellow and dark green

Unique Characteristics: Foliage is banded or striped yellow; can be curled to create unusual design elements.

Design Applications: (large) (mass) Bold leaf provides mass and interest. Good candidate to use with tropical flowers or exotics.

Longevity: weeks 1-2

Genus Name (Latin): *Anthurium*

Pronunciation: (an-THEW-ree-um)

Species Name (Latin): spp.

Common Name: **Anthurium**

Family Name (Latin and Common): *Araceae,* the Arum family

Related Family Members in Book: *Caladium, Monstera, Philodendron,* and *Zantedeschia*

Availability: indoor plant: year-round

Color: green (dark)

Unique Characteristics: Foliage of tropical plant has shiny, almost waxy appearance; displays a unique form.

Design Applications: (medium) (form) Sturdy leaf provides visual mass. *Anthurium* leaves work best with other exotic or tropical flowers.

Longevity: weeks 1-2

Genus Name (Latin): ***Artemisia***

Pronunciation: (ar-te-MIZ-ee-a; ar-te-MISH-ee-a)

Species Name (Latin): sp.

Common Name: ***Artemisia,*** Mugwort

Family Name (Latin and Common): *Asteraceae,* the Composite or Sunflower family

Related Family Members in Book: *Achillea, Ageratum, Aster, Bracteantha, Centaurea, Chrysanthemum, Cosmos, Craspedia, Dahlia, Echinacea, Gerbera, Helianthus, Leucanthemum, Liatris, Rudbeckia, Senecio, Solidago, x Solidaster,* and *Zinnia*

Availability: commercial product: July-October

Color: gray

Unique Characteristics: Air-dries well; is suitable for pressing.

Design Applications: (small) (filler) Leaf is uniquely gray. Makes an excellent filler for mixed flower bouquets, and is especially beautiful in combination with bright-colors. Good candidate for pressed floral creations.

Longevity: days 6-10

Genus Name (Latin): ***Asparagus***

Pronunciation: (a-SPAR-a-gus)

Species Name (Latin): ***asparaginoides***

Common Name: **String Smilax,** Greenbriar

Family Name (Latin and Common): *Liliaceae,* the Lily family

Related Family Members in Book: *Aspidistra, Gloriosa, Hemerocallis, Hosta, Hyacinthus, Kniphofia, Liriope, Muscari, Ornithogalum, Ruscus, Sandersonia, Scilla, Smilax, Tulipa,* and *Xerophyllum*

Availability: commercial product: year-round

Color: green (dark)

Unique Characteristics: Foliage is dark green with lanceolate leaves. Commercially available grown along a piece of string to produce a garland approximately 3-4 inches in diameter.

Design Applications: (medium) (line) Soft appearing garland is useful for bridal designs and decorations.

Longevity: days 7-10, if stored inside airtight plastic bag before use.

Genus Name (Latin): ***Asparagus***   Pronunciation: (a-SPAR-a-gus)

Species Name (Latin): ***densiflorus* 'Myersii'**
syn. *A. meyeri, A. 'Myers'*

Common Name: **Foxtail Fern**

Family Name (Latin and Common): *Liliaceae,*
the Lily Family

Related Family Members in Book: *Aspidistra,
Gloriosa, Hemerocallis, Hosta, Hyacinthus,
Kniphofia, Liriope, Muscari, Ornithogalum,
Ruscus, Sandersonia, Scilla, Smilax, Tulipa,* and
Xeropyllum.

Availability: commercial product: year-round

Color: green (medium)

Unique Characteristics: Fronds are foxtail-like
being stiffly erect, and very densely covered.

Design Applications: (small) (line) This foliage
is nice for use as a line element or for adding
texture. Use sparingly.

Longevity: weeks 1-2

Genus Name (Latin): ***Asparagus***   Pronunciation: (a-SPAR-a-gus)

Species Name (Latin): ***densiflorus* 'Sprengeri'**

Common Name: **Sprengeri,** Sprenger Asparagus

Family Name (Latin and Common): *Liliaceae,*
the Lily family

Related Family Members in Book: *Aspidistra,
Gloriosa, Hemerocallis, Hosta, Hyacinthus,
Kniphofia, Liriope, Muscari, Ornithogalum,
Ruscus, Sandersonia, Scilla, Smilax, Tulipa,* and
Xerophyllum

Availability: commercial product: year-round

Color: green (medium)

Unique Characteristics: Airy, often trailing
stems; thorny; sometimes displays green
berries; newer growth appears light green.

Design Applications: (medium) (filler) Linear
stem of this cut foliage is available in long
lengths, making it a popular choice for
producing garlands for bridal and party events.
Short lateral stems are useful in corsage and
boutonniere designs. The plant is also used

extensively in hanging baskets, window boxes,
and urns, as a durable filler.

Longevity: weeks 1-2

Genus Name (Latin): ***Asparagus***   Pronunciation: (a-SPAR-a-gus)

Species Name (Latin): ***pyramidalis***

Common Name: **Tree Fern**

Family Name (Latin and Common): *Liliaceae,* the Lily family

Related Family Members in Book: *Aspidistra, Gloriosa, Hemerocallis, Hosta, Hyacinthus, Kniphofia, Liriope, Muscari, Ornithogalum, Ruscus, Sandersonia, Scilla, Smilax, Tulipa,* and *Xerophyllum*

Availability: commercial product: year round

Color: green (dark)

Unique Characteristics: Foliage is airy, soft in appearance and touch; can give "hairy" look to designs if overused.

Design Applications: (small) (filler, texture) Feathery foliage adds a fine texture, and is commonly used in body flower designs.

Longevity: weeks 1-2

Genus Name (Latin): ***Asparagus***   Pronunciation: (a-SPAR-a-gus)

Species Name (Latin): ***setaceus*** syn. *A. plumosus*

Common Name: **Plumosa Fern,** Asparagus Fern

Family Name (Latin and Common): *Liliaceae,* the Lily family

Related Family Members in Book: *Aspidistra, Gloriosa, Hemerocallis, Hosta, Hyacinthus, Kniphofia, Liriope, Muscari, Ornithogalum, Ruscus, Sandersonia, Scilla, Smilax, Tulipa,* and *Xerophyllum*

Availability: commercial product: year-round

Color: green (medium to dark)

Unique Characteristics: Foliage has airy, soft appearance; flat form can be pressed; sometimes can appear "hairy" in designs if overused.

Design Applications: (small) (filler, texture) Feathery foliage has graceful stems; delicate-looking fern adds delightful texture to any style. Used extensively for bridal, corsage, and boutonniere work, and is a nice material for

creating garlands. A good candidate for pressed flower designs.

Longevity: weeks 1-2

Genus Name (Latin): ***Asparagus***   Pronunciation: (a-SPAR-a-gus)

Species Name (Latin): spp.

Common Name: **Ming Fern**

Family Name (Latin and Common): *Liliaceae,* the Lily family

Related Family Members in Book: *Aspidistra, Gloriosa, Hemerocallis, Hosta, Hyacinthus, Kniphofia, Liriope, Muscari, Ornithogalum, Ruscus, Sandersonia, Scilla, Smilax, Tulipa,* and *Xerophyllum*

Availability: commercial product: year-round

Color: green (light to medium)

Unique Characteristics: Needlelike leaves are short and soft.

Design Applications: (small) (filler) Short pieces can be used for corsage and boutonniere work. If overused, can become visually heavy or "fuzzy" in appearance.

Longevity: weeks 1-2

Genus Name (Latin): ***Aspidistra***   Pronunciation: (as-pi-DIS-tra)

Species Name (Latin): ***elatior***

Common Name: **Aspidistra,** Cast-Iron Plant, Barroom Plant

Family Name (Latin and Common): *Liliaceae,* the Lily family

Related Family Members in Book: *Asparagus, Gloriosa, Hemerocallis, Hosta, Hyacinthus, Kniphofia, Liriope, Muscari, Ornithogalum, Ruscus, Sandersonia, Scilla, Smilax, Tulipa,* and *Xerophyllum*

Availability: commercial product: year-round with supplies limited in winter months

Color: green (dark)

Unique Characteristics: Extremely long lasting foliage has stiff stem and leaf; appearance is similar to *Cordyline terminalis;* can be altered to create an abstract shape.

Design Applications: (large) (mass) Linear foliage always adds mass. Great in contemporary designs, especially in combination with tropical flowers.

Longevity: weeks 3-4 long lasting

Genus Name (Latin): ***Aspidistra***

Pronunciation: (as-pi-DIS-tra)

Species Name (Latin): ***elatior* 'Variegata'**

Common Name: ***Aspidistra,*** Cast-Iron Plant, Barroom Plant

Family Name (Latin and Common): *Liliaceae,* the Lily family

Related Family Members in Book: *Asparagus, Gloriosa, Hemeorcallis, Hosta, Hyacinthus, Kniphofia, Liriope, Muscari, Ornithogalum, Ruscus, Sandersonia, Scilla, Smilax, Tulipa,* and *Xerophyllum*

Availability: commercial product: year-round with limited supplies in winter months

Color: green with varied white striping

Unique Characteristics: Extremely long lasting foliage has stiff stem and leaf; appearance is similar to *Cordyline terminalis;* can be altered to create abstract shapes.

Design Applications: (large) (mass) Linear foliage is great for use in contemporary

designs, especially in combination with tropical flowers.

Longevity: weeks 3-4 long lasting

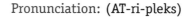

Genus Name (Latin): ***Atriplex***

Pronunciation: (AT-ri-pleks)

Species Name (Latin): spp.

Common Name: ***Atriplex***

Family Name (Latin and Common): *Chenopodiaceae,* the Goosefoot family

Related Family Members in Book: none

Availability: commercial product: year-round, varied supply in late summer

Color: green (shown), burgundy red

Unique Characteristics: Color is wonderful; several available.

Design Applications: (large) (texture) Linear material is an excellent choice for adding texture and color. Useful as a filler and sometimes as a line element.

Longevity: days 7-10

Genus Name (Latin): ***Aucuba***   Pronunciation: (ah-KEW-ba)

Species Name (Latin): *japonica* **'Crotonifolia'**

Common Name: ***Aucuba,*** Japanese Laurel

Family Name (Latin and Common): *Cornaceae,* the Dogwood family

Related Family Members in Book: *Cornus*

Availability: commercial product: year-round

Color: green with yellow speckles

Unique Characteristics: Stem is woody, holding varying sizes of glossy leaves.

Design Applications: (large) (accent) Leaf is an excellent choice for adding accent. Useful as a foundation or background foliage, always adding mass.

Longevity: days 7-10

Genus Name (Latin): ***Avena***   Pronunciation: (a-VEE-na)

Species Name (Latin): sp.

Common Name: **Avena Grass**

Family Name (Latin and Common): *Poaceae,* the Grass family

Related Family Members in Book: *Chasmanthium, Eleusine, Miscanthus, Pennisetum, Phalaris, Phragmites, Phyllostachys, Setaria, Sorghum,* and *Triticum*

Availability: commercial product: year-round

Color: tan

Unique Characteristics: Dried material; seed head is tiny.

Design Applications: (small) (texture) This multi-stemmed dried material is an excellent choice for adding texture and movement to permanent wall décor, centerpieces, or vase arrangements.

Longevity: long lasting

Genus Name (Latin): ***Betula***   Pronunciation: (BET-yew-la)

Species Name (Latin): spp.

Common Name: **Birch**

Family Name (Latin and Common): *Betulaceae*, the Birch family

Related Family Members in Book: none

Availability: commercial product: year-round

Color: brown

Unique Characteristics: Dried material; stem has catkins.

Design Applications: (large) (line) Branch material is an excellent line element for perishable and nonperishable floral designs. Catkins add interest. Also used for framework or armatures.

Longevity: long lasting

Genus Name (Latin): ***Bracteantha***   Pronunciation: (brak-tee-ANTH-a)

Species Name (Latin): ***bracteata*** cultivar syn. *Helichrysum bracteatum*

Common Name: **Strawflower**

Family Name (Latin and Common): *Asteraceae*, the Composite or Sunflower family

Related Family Members in Book: *Achillea, Ageratum, Artemisia, Aster, Centaurea, Chrysanthemum, Cosmos, Craspedia, Dahlia, Echinops, Echinacea, Gerbera, Helianthus, Leucanthemum, Liatris, Rudbeckia, Senecio, Solidago, x Solidaster,* and *Zinnia*

Availability: commercial product: varies August-November

Color: orange, yellow, bronze, pink, white, and red

Unique Characteristics: Flower has wonderful papery appearance, and strawlike texture; dried flower is commercially available on wire stems; periodically appears as a perishable product that can be air-dried; many cultivars available.

Design Applications: (small) (mass flower) Round dried flower provides texture and mass.

Longevity: long lasting

Genus Name (Latin): ***Buxus*** Pronunciation: (BUK-sus)

Species Name (Latin): spp.

Common Name: **Boxwood,** Box

Family Name (Latin and Common): *Buxaceae,* the Box family

Related Family Members in Book: *Pachysandra*

Availability: commercial product: year-round

Color: green (dark)

Unique Characteristics: Stem is woody with stiff leaves like oregonia; foliage is very durable.

Design Applications: (small) (texture) Tiny leaf is a nice choice for providing textural interest to traditional and contemporary styles. Short lateral stems are excellent filler.

Longevity: weeks 1-2

Genus Name (Latin): ***Buxus*** Pronunciatio n: (BUK-sus)

Species Name (Latin): spp.

Common Name: **Oregonia**

Family Name (Latin and Common): *Buxaceae,* the Box family

Related Family Members in Book: *Pachysandra*

Availability: commercial product: year-round

Color: green with white margins

Unique Characteristics: Stem and leaf are stiff; foliage is very durable.

Design Applications: (small) (texture) Tiny variegated leaf is a nice choice for textural interest as a filler foliage. Lateral stems are useful in body flower designs, for example, boutonnieres and corsages.

Longevity: weeks 1-2

Genus Name (Latin): **Caladium**   Pronuncation: (ka-LAY-dee-um)

Species Name (Latin): **x *Hortulanum*** cultivar

Common Name (Latin): ***Caladium,*** Fancy-
Leaved Caladium, Angel Wings, Mother-in-Law
Plant

Family Name (Latin and Common): *Araceae,* the
Arum family

Related Family Members in Book: *Anthurium,
Monstera, Philodendron,* and *Zantedeschia*

Availability: landscape plant: year-round

Color: red, rose, salmon, and white variations

Unique Characteristics: Leaf can be flat,
undulate, or ruffled; used often to lighten up
shaded portions of temporary landscape
displays; appears ovate to lanceolate in shape
with many variations of color in cultivars.

Design Applications: (large) (form) Single leaf
adds mass to cut flower designs. The plant also
works well in containerized gardens or
temporary landscapes.

Longevity: days 3-4, as a cut foliage

Genus Name (Latin): **Calathea**   Pronunciation: (kal-a-THEE-a)

Species Name (Latin): spp.

Common Name: **Calathea,** Peacock Plant

Family Name (Latin and Common): *Marantaceae,*
the Maranta or Arrowroot family

Related Family Members in Book: none

Availability: commercial product: year-round

Color: green

Unique Characteristics: Leaf is dark green,
showing elliptic to oblong spots along the
lateral veins.

Design Applications: (medium) (mass) Nice
foliage has distinctive venation which always
draws attention. Best uses are with tropical
flowers in stylized arrangements, which
highlight its unique characteristic.

Longevity: weeks 1-2

Genus Name (Latin): ***Callicladium***	Pronunciation: (kal-i-CLAY-dee-um)

Species Name (Latin): ***haldanianum***

Common Name: **Florist's Sheet Moss**

Family Name (Latin and Common): *Hypnaceae,* the Feather Moss family

Related Family Members in Book: none

Availability: commercial product: year-round

Color: green shades

Unique Characteristics: Dried material; commonly used as sheet moss, is actually part of several genera. Listed here because of its extensive use, and does vary by species; photograph shows dried specimen, but these mosses are often available in a perishable form.

Design Applications: (small) (texture) Material is a florist's staple. Moss is useful to amateurs and professionals to hide mechanics of construction.

Longevity: long lasting

Genus Name (Latin): ***Callistemon***	Pronunciation: (kal-i-STEE-mon)

Species Name (Latin): spp.

Common Name: **Bottlebrush**

Family Name (Latin and Common): *Myrtaceae,* the Myrtle family

Related Family Members in Book: *Callistemon, Eucalyptus, Leptospermum, Melaleuca, Myrtus,* and *Thryptomene*

Availability: commercial product: year-round

Color: green (light)

Unique Characteristics: Foliage is commercially sold but most often contains some red flowers, inflorescence of long, exserted stamens resembles a bottlebrush.

Design Applications: (small) (filler) Long, linear stem is a nice product for use as a filler or accent. Makes a nice background foliage for smaller-scaled designs.

Longevity: days 7-10

Genus Name (Latin): *Calocedrus*   Pronunciation: (cal-oh-SED-rus)

Species Name (Latin and Common): *decurrens*

Common Name: **Incense Cedar**

Family Name (Latin and Common): *Cupressaceae,* the Cypress family

Related Family Members in Book: *Juniperus*

Availability: commercial product: November-January

Color: green (medium)

Unique Characteristics: Displays stems of flat sprays with yellow accents.

Design Applications: (large) (line) Flat sprays, often with long lateral branches, are useful for adding line to oblong centerpieces. Different texture is a nice addition to designs of mixed conifers. Useful for any design style or application in holiday decorations.

Longevity: weeks 3-4 long lasting

Genus Name (Latin): *Camellia*   Pronunciation: (ka-MEEL-ee-a; ka-MEEL-ya)

Species Name (Latin): spp.

Common Name: *Camellia*

Family Name (Latin and Common): *Theaceae,* the Tea family

Related Family Members in Book: none

Availability: commercial product: year-round

Color: green (dark)

Unique Characteristics: Stem is woody with glossy leaf; durable.

Design Applications: (large) (mass) Leaves on heavy stems add mass. A good background material. Smaller leaves are often used in body flower designs, for instance, corsages and boutonnieres.

Longevity: days 5-7

Genus Name (Latin): ***Catalpa***   Pronunciation: (ka-TAL-pa)

Species Name (Latin): ***bignonioides***

Common Name: **Southern Catalpa**, Common Catalpa, Indian Bean

Family Name (Latin and Common): *Bignoniaceae,* the Bignonia family

Related Family Members in Book: none

Availability: landscape plant: typically not available as a commercial dried material

Color: brown (dark)

Unique Characteristics: Seed pod is pendulous; typically 8-20 inches long.

Design Applications: (medium) (texture) Pod has knobby appearance and provides texture to dried wreaths.

Longevity: varies

Genus Name (Latin): ***Chamaecyparis***   Pronunciation: (kam-e-SIP-a-ris)

Species Name (Latin): ***lawsoniana***

Common Name: **Port Orford Cedar**, Lawson False Cypress

Family Name (Latin and Common): *Cupressaceae,* the Cypress family

Related Family Members in Book: *Juniperus*

Availability: commercial product: November-January

Color: green (light)

Unique Characteristics: Stem has flat sprays with scalelike overlapping leaves; also available preserved with glycerine.

Design Applications: (large) (filler) Flat filler is an excellent choice for adding length. Beautiful when used in combination with other conifers of varying texture and shades of color. Adapts well to any kind of holiday décor.

Longevity: weeks 3-4 long lasting

Genus Name (Latin): ***Chamaecyparis*** Pronunciation: (kam-e-SIP-a-ris)

Species Name (Latin): *pisifera* **'Filifera Aurea'**

Common Name: **Sawara Cypress**

Family Name (Latin and Common): *Cupressaceae,* the Cypress family

Related Family Members in Book: *Juniperus*

Availability: landscape plant: not typically available as a commercial cut foliage.

Color: green (yellow)

Unique Characteristics: Has stem of flattened sprays; with scale-like leaves; golden yellow leaves are distinctive.

Design Applications: (medium) (accent) Golden-tipped conifer is most effective when used in small amounts among mixed conifers in holiday designs.

Longevity: weeks 1-3

Genus Name (Latin): ***Chamaedorea*** Pronunciation: (kam-a-DO-ree-a)

Species Name (Latin): spp.

Common Name: **Florists' Emerald**

Family Name (Latin and Common): *Arecaceae,* the Palm family

Related Family Members in Book: *Cocos, Sabal*

Availability: commercial product: year-round

Color: green (medium)

Unique Characteristics: Form is adaptable for braiding or cutting into geometric shapes; plant is the same genus as florists' jade, which is also heavily used in large arrangements.

Design Applications: (large) (filler) Frond is very versatile, extensively used for framing and to cover mechanics; can also be used to add accent when cut or braided. Favored choice of the florist for use in traditional line mass designs of large size.

Longevity: weeks 2-3 long lasting

Genus Name (Latin): ***Chamaedorea*** Pronunciation: (kam-a-DO-ree-a)

Species Name (Latin): spp.

Common Name: **Florists' Jade**

Family Name (Latin and Common): *Arecaceae,* the Palm family

Related Family Members in Book: *Cocos, Sabal*

Availability: commercial product: year-round

Color: green (medium)

Unique Characteristics: Form is adaptable to cutting into geometric shapes.

Design Applications: (large) (filler) Leaf is commonly used for framing and covering mechanics, and is a staple for the florist who produces large-scaled line mass designs. Useful for creating abstract-shaped leaves for contemporary styles.

Longevity: weeks 2-3 long lasting

Genus Name (Latin): ***Chasmanthium*** Pronunciation: (chas-MAN-thee-um)

Species Name (Latin): ***latifolium***

Common Name: **River Oats,** Inland Sea Oats

Family Name (Latin and Common): *Poaceae,* the Grass family

Related Family Members in Book: *Avena, Eleusine, Miscanthus, Pennisetum, Phalaris, Phragmites, Phyllostachys, Setaria, Sorghum,* and *Triticum*

Availability: landscape plant: limited availability as a cut foliage

Color: green (medium to light)

Unique Characteristics: Air-dries well; seed head is a unique light green color, changing to tan in autumn; periodically found in the retail market.

Design Applications: (small) (filler) Seed head provides movement and can be used to add accent or as a filler. Makes a nice choice in combination with exotics and other grasses.

Longevity: days 5-7

Genus Name (Latin): ***Chlorophytum***   Pronunciation: (klo-row-FI-tum)

Species Name (Latin): ***comosum* 'Vittatum'**

Common Name: **Spider Plant**, Spider Ivy, Ribbon Plant

Family Name (Latin and Common): *Liliaceae,* the Lily family

Related Family Members in Book: *Asparagus, Aspidistra, Gloriosa, Hemerocallis, Hosta, Hyacinthus, Muscari, Ornithogalum, Ruscus, Sandersonia, Scilla, Tulipa,* and *Xerophyllum*

Availability: indoor plant: year-round

Color: green with central white stripe

Unique Characteristics: Plant has plantlets that are spidery in appearance.

Design Applications: (small) (accent) Plantlets provide movement and interest to handheld bouquets. Bold striping is a nice accent to mixed container gardens.

Longevity: varies

Genus Name (Latin): ***Cladina***   Pronunciation: (kla-DY-na)

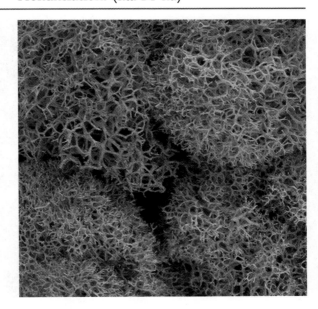

Species Name (Latin): sp.

Common Name: **Florists' Reindeer Moss**

Family Name (Latin and Common): *Cladinaceae,* the Reindeer Lichen family

Related Family Members in Book: none

Availability: commercial product: year-round

Color: gray

Unique Characteristics: Dried material has a rounded form.

Design Applications: (small) (accent, texture) Material is good for adding basal interest to designs. Useful for covering mechanics and pillowing techniques.

Longevity: long lasting

Genus Name (Latin): ***Cocos***   Pronunciation: (KO-kos)

Species Name (Latin): ***nucifera***

Common Name: **Coco Lashing**

Family Name (Latin and Common): *Arecaceae,* the Palm family

Related Family Members in Book: *Chamaedorea, Sabal*

Availability: commercial product: year-round

Color: brown (light)

Unique Characteristics: Dried material comes from the edible coconut plant; is also available in other dyed colors.

Design Applications: (large) (line) This dried material is an excellent choice for an interesting line element. Use in contemporary permanent arrangements.

Longevity: long lasting

Genus Name (Latin): ***Codiaeum***   Pronunciation: (co-dee-AY-um)

Species Name (Latin): ***variegatum*** cultivar

Common Name: **Croton**, Variegated Laurel

Family Name (Latin and Common): *Euphorbiaceae,* the Euphorbia family

Related Family Members in Book: *Acalypha, Euphorbia,* and *Sapium*

Availability: commercial product: year-round

Color: red, yellow, green, orange, often spotted and blotched irregularly

Unique Characteristics: Colors are bold; not to be confused with the genus *Croton;* several cultivars available, some having twisted leaves.

Design Applications: (medium-large) (accent) Distinctive leaf demands attention. A good choice for tropical flower and foliage combinations. Best use is for contemporary styles.

Longevity: days 4-6

Genus Name (Latin): ***Cordyline*** Pronunciation: (kor-di-LY-nee)

Species Name (Latin): ***fruticosa*** cultivar syn. *Dracaena terminalis*

Common Name: **Ti Tree,** Good-Luck Tree, Tree of Kings

Family Name (Latin and Common): *Agavaceae,* the Agave family

Related Family Members in Book: *Dracaena, Phormium, Polianthes,* and *Yucca*

Availability: commercial product: year-round

Color: green

Unique Characteristics: Size is often 5 inches by 1 foot long; some cultivars are available with pinkish-red margins. Flowers can be curled or cut into shapes.

Design Applications: (large) (mass) Leaf is best suited for contemporary styles, especially combinations of tropicals.

Longevity: weeks 1-2

Genus Name (Latin): ***Cotinus*** Pronunciation: (KOT-i-nus; ko-TY-nus)

Species Name (Latin): ***coggygria* 'Velvet Cloak'**

Common Name: **Common Smoketree,** Common Smokebush

Family Name (Latin and Common): *Anacardiaceae,* the Cashew family

Related Family Members in Book: *Rhus, Schinus*

Availability: landscape plant: commercially limited as a cut foliage.

Color: purple

Unique Characteristics: Color is a handsome dark purple, turning rich reddish purple in fall; not regularly available as a commercial cut foliage, but periodically appears; occasionally available with wonderful smokelike plumes.

Design Applications: (large) (accent) Unique color and oval leaf form add mass and accent. Useful for Ikebana designs.

Longevity: days 3-5 or longer

Genus Name (Latin): ***Crocosmia***   Pronunciation: (kro-KOZ-mi-a)

Species Name (Latin): **x crocosmiiflora**

Common Name: **Montbretia Pods**

Family Name (Latin and Common): *Iridaceae,* the Iris family

Related Family Members in Book: *Freesia, Gladiolus, Iris, Ixia,* and *Watsonia*

Availability: commercial product: August-November

Color: brown, green to orange

Unique Characteristics: Air-dries well; typically available as a perishable product; coloration varies.

Design Applications: (small) (texture) Seed capsules along linear stem are good additions of texture to any style. Works well in autumnal designs—both contemporary and traditional, perishable and dried.

Longevity: long lasting

Genus Name (Latin): ***Cycas***   Pronunciation: (SY-kus)

Species Name (Latin): **revoluta**

Common Name: **Cycas Palm,** Sago Palm

Family Name (Latin and Common): *Cycadaceae,* the Cycas or Cycad family

Related Family Members in Book: none

Availability: commercial product: year-round

Color: green (dark)

Unique Characteristics: Frond is about 20 inches long with stiff needles, sharp tips, and is extremely durable.

Design Applications: (large) (accent, texture, mass) Dramatic leaf of incredible size is perfect for large tropical arrangements, for instance, lobby or buffet pieces.

Longevity: weeks 2-3 long lasting

Genus Name (Latin): ***Cyclamen***   Pronunciation: (SIK-la-men; SY-kla-men)

Species Name (Latin): spp.

Common Name: ***Cyclamen***

Family Name (Latin and Common): *Primulaceae,* the Primrose family

Related Family Members in Book: *Lysimachia*

Availability: indoor plant: year-round

Color: green

Unique Characteristics: Form of leaf is heart shaped, often with gray-green marbling.

Design Applications: (medium) (form) Leaf works well in contemporary bridal bouquets. Smaller leaves are suitable for boutonnieres, corsages, and container arrangements.

Longevity: days 3-4, as a cut foliage

Genus Name (Latin): ***Cynara***   Pronunciation: (SIN-a-ra)

Species Name (Latin): sp.

Common Name: **Artichoke,** Chico Choke

Family Name (Latin and Common): *Asteraceae,* the Composite or Sunflower family

Related Family Members in Book: *Achillea, Ageratum, Artemisia, Aster, Bracteantha, Centaurea, Chrysanthemum, Cosmos, Craspedia, Dahlia, Echinacea, Echinops, Gerbera, Helianthus, Leucanthemum, Liatris, Rudbeckia, Senecio, Solidago, x Solidaster,* and *Zinnia*

Availability: commercial product: year-round

Color: tan

Unique Characteristics: Dried material has a distinctive form; often available on wire stems as shown in photo; also commercially available in painted colors.

Design Applications: (large) (mass) Large, rounded form provides mass to dried bouquets.

Longevity: long lasting

Genus Name (Latin): **Cyperus**

Pronunciation: (sy-PEE-rus; si-PEE-rus)

Species Name (Latin): *alternifolius*

Common Name: **Umbrella Palm,** Umbrella Plant, Umbrella Sedge

Family Name (Latin and Common): *Cyperaceae,* the Sedge family

Related Family Members in Book: none

Availability: commercial product: year-round

Color: green (medium)

Unique Characteristics: Stem is leafless with terminal whorl of leaves; flower spikes are sometimes evident; also available preserved with glycerine.

Design Applications: (medium) (accent) Green "umbrella" is interesting. Good choice for use with other exotic and tropical flowers. Best suited for contemporary designs to show off distinct form.

Longevity: weeks 1-3

Genus Name (Latin): **Cyperus**

Pronunciation: (sy-PEE-rus; si-PEE-rus)

Species Name (Latin): *papyrus*

Common Name: **Lion's Head Papyrus,** Bulrush, Paper Plant

Family Name (Latin and Common): *Cyperaceae,* the Sedge family

Related Family Members in Book: none

Availability: commercial product: year-round

Color: green (medium)

Unique Characteristics: Stem is leafless; displays shorter tufts and thinner rays than *C. alternifolius;* ancient Egyptians flattened and dried the stems to make a form of paper.

Design Applications: (large) (accent). Very unusual form adds drama. Best used in high-style designs in combination with exotic or tropical flowers.

Longevity: weeks 1-3

Genus Name (Latin): ***Cyrtomium***   Pronunciation: (sur-TO-mee-um)

Species Name (Latin): ***falcatum***

Common Name: **Holly Fern**

Family Name (Latin and Common): *Polypodiaceae,* the Polypody family

Related Family Members in Book: *Adiantum, Nephrolepsis, Platyactium, Polystichum,* and *Platycerium*

Availability: indoor plant: year-round

Color: green (dark)

Unique Characteristics: Leaf is glossy, sometimes with margins coarsely fringed or deeply serrated.

Design Applications: (large) (accent) Glossy leaf is useful as an accent. Nice addition to high-style designs.

Longevity: days 6-10

Genus Name (Latin): ***Cytisus***   Pronunciation: (SIT-is-us)

Species Name (Latin): ***scoparius***

Common Name: **Scotch Broom,** Broom

Family Name (Latin and Common): *Fabaceae,* the Pea or Pulse family

Related Family Members in Book: *Acacia, Cercis, Lathyrus, Lupinus,* and *Wisteria*

Availability: commercial product: August-April

Color: green (dark)

Unique Characteristics: Stem is woody with many slender branchlets; very durable; photo shows flowering stage.

Design Applications: (large) (line) Large stem creates strong lines good for framework. Slender branchlets are suitable for line material in small arrangements. Can be curved or bent to produce visual tension. Suitable for adding height or length to contemporary and traditional designs.

Longevity: weeks 3 long lasting

Genus Name (Latin): ***Dracaena*** Pronunciation: (dra-SEE-na)

Species Name (Latin): *fragrans* **'Warneckei'**

Common Name: **Striped Dracaena**

Family Name (Latin and Common): *Agavaceae,* the Agave family

Related Family Members in Book: *Cordyline, Phormium, Polianthes,* and *Yucca*

Availability: commercial product: year-round

Color: green with white stripes

Unique Characteristics: Slender variegated leaf; several *Dracaena* species are available as potted houseplants and cut foliage.

Design Applications: (small) (line) Unique leaf provides interesting line. Useful in tropical arrangements and contemporary bridal or attendants' bouquets.

Longevity: weeks 2-4

Genus Name (Latin): ***Eleusine*** Pronunciation: (el-you-SY-nee)

Species Name (Latin): ***corcana***

Common Name: **Finger Millet,** African Millet

Family Name (Latin and Common): *Poaceae,* the Grass family

Related Family Members in Book: *Avena, Chasmanthium, Miscanthus, Pennisetum, Phalaris, Phragmites, Phyllostachys, Setaria, Sorghum,* and *Triticum*

Availability: commercial product: year-round

Color: green (basil), tan (natural)

Unique Characteristics: Dried material; seed spikes have the appearance of a finger.

Design Applications: (medium) (texture) Spike produces an excellent linear texture. Works well in bundling techniques. Nice when grouped with other dried materials of mixed forms.

Longevity: long lasting

Genus Name (Latin): ***Eleusine***　　Pronunciation: (el-you-SY-nee)

Species Name (Latin): sp.

Common Name: **Tapestry Millet**

Family Name (Latin and Common): *Poaceae,* the Grass family

Related Family Members in Book: *Avena, Chasmanthium, Miscanthus, Pennisetum, Phalaris, Phragmites, Phyllostachys, Setaria, Sorghum,* and *Triticum*

Availability: commercial product: year-round

Color: green (basil), tan (natural)

Unique Characteristics: Dried material; wonderful spike-like seed head.

Design Applications: (medium) (texture) Seed head produces an excellent linear texture. Beautiful when used with other dried flowers and grasses in home décor. A nice addition to autumnal cut-flower centerpieces.

Longevity: long lasting

Genus Name (Latin): ***Equisetum***　　Pronunciation: (ek-wi-SEE-tum)

Species Name (Latin): ***hyemale***

Common Name: **Horsetail,** Snake Grass, Scouring Rush

Family Name (Latin and Common): *Equisetaceae,* the Horsetail family

Related Family Members in Book: none

Availability: commercial product: year-round

Color: green

Unique Characteristics: Stem is furrowed with varying sized rough ridges; conelike spikes at top contain spores; leafless, aquatic plant is found in lowlands, including pool margins; has hollow stem.

Design Applications: (small) (line) Versatile material is well suited for contemporary designs. Fantastic choice for creating strong angled lines in geometric or abstract arrangements.

Longevity: weeks 1-3

Genus Name (Latin): *Eucalyptus*

Pronunciation: (yew-ka-LIP-tus)

Species Name (Latin): *coccifera*

Common Name: **Flowering Euc,** Mount Wellington Peppermint, Tasmanian Snow Gum

Family Name (Latin and Common): *Myrtaceae,* the Myrtle family

Related Family Members in Book: *Callistemon, Chamelaucium, Leptospermum, Lophomyrtus, Melaleuca, Myrtus,* and *Thryptomene*

Availability: commercial product: varies year-round

Color: gray-green

Unique Characteristics: Seed is distinctive; adult leaves are peppermint scented. Flowers shown are not typically present—leaf form visible is indicative of what is sold commercially; flowers do not last as long as foliage.

Design Applications: (small) (filler) Elliptic leaves with unusual flowers add accent. Works well with other exotics. Highlight the flower when present.

Longevity: weeks 1-2, for leaves

Genus Name (Latin): *Eucalyptus*

Pronunciation: (yew-ka-LIP-tus)

Species Name (Latin): *gunnii*

Common Name: **Cider Gum,** Silver Spoon Eucalyptus

Family Name (Latin and Common): *Myrtaceae,* the Myrtle family

Related Family Members in Book: *Callistemon, Chamelaucium, Leptospermum, Lophomyrtus, Melaleuca, Myrtus,* and *Thryptomene*

Availability: commercial product: year-round

Color: gray-brown

Unique Characteristics: Fragrance is pungent; air-dries well, becoming brittle.

Design Applications: (small) (filler) A useful filler for adding unusual color and texture.

Longevity: weeks 2-3 long lasting

Genus Name (Latin): *Eucalyptus*   Pronunciation: (yew-ka-LIP-tus)

Species Name (Latin): *nicholii*

Common Name: **Willow Eucalyptus,** Nichol's Willow-Leaved Peppermint, Narrow-Leaved Black Peppermint

Family Name (Latin and Common): *Myrtaceae,* the Myrtle family

Related Family Members in Book: *Callistemon, Chamelaucium, Leptospermum, Lophomyrtus, Melaleuca, Myrtus,* and *Thryptomene*

Availability: commercial product: year-round

Color: gray-green

Unique Characteristics: Fragrance is pungent; narrow-lanceolate leaf shape; will air-dry, becoming brittle; many *Eucalyptus* available, often containing berries.

Design Applications: (small) (filler) Distinctive filler provides great texture for any style.

Longevity: weeks 2-3 long lasting

Genus Name (Latin): *Eucalyptus*   Pronunciation: (yew-ka-LIP-tus)

Species Name (Latin): *nicholii*

Common Name: **Willow Eucalyptus,** Nichol's Willow-Leaved Peppermint, Narrow-Leaved Black Peppermint

Family Name (Latin and Common): *Myrtaceae,* the Myrtle family

Related Family Members in Book: *Callistemon, Chamelaucium, Leptospermum, Lophomyrtus, Melaleuca, Myrtus,* and *Thryptomene*

Availability: commercial product: year-round

Color: red (shown), green, blue, and brown

Unique Characteristics: Dried material; photo shows example of preserved and dyed material that is commercially available.

Design Applications: (small) (filler) Leaf material is used extensively in permanent home décor because of its fragrance and unique form. Provides a great texture for any style of arrangement.

Longevity: long lasting

Genus Name (Latin): *Eucalyptus*   Pronunciation: (yew-ka-LIP-tus)

Species Name (Latin): *polyanthemos*

Common Name: **Silver-Dollar Eucalyptus,** Silver-Dollar Gum, Silver-Dollar Tree

Family Name (Latin and Common): *Myrtaceae,* the Myrtle family

Related Family Members in Book: *Callistemon, Chamelaucium, Leptospermum, Lophomyrtus, Melaleuca, Myrtus,* and *Thryptomene*

Availability: commercial product: year-round

Color: gray-green

Unique Characteristics: Fragrance is pungent; leaves are orbicular to broadly lanceolate, giving silver-dollar appearance; will air dry, becoming brittle.

Design Applications: (small) (filler) Adds textural interest with leaf form and berries. Makes a nice choice for mixed bouquets as a filler, or to produce emphasis in focal zones by adding texture.

Longevity: weeks 2-3 long lasting

Genus Name (Latin): *Eucalyptus*   Pronunciation: (yew-ka-LIP-tus)

Species Name (Latin): *pulverulenta*

Common Name: *Eucalyptus*

Family Name (Latin and Common): *Myrtaceae,* the Myrtle family

Related Family Members in Book: *Callistemon, Chamelaucium, Leptospermum, Lophomyrtus, Melaleuca, Myrtus,* and *Thryptomene*

Availability: commercial product: year-round

Color: gray, silvery-blue

Unique Characteristics: Fragrance is strong; commercially available dyed and treated with glycerine for pliability; popular in permanent botanical designs; preserved *Eucalyptus* is available in several colors.

Design Applications: (medium) (filler) Stem length and form give this filler versatility, making it adaptable to large or small designs. Unique form adds interest when used for line, filler, or texture.

Longevity: weeks 2-3 long lasting

Genus Name (Latin): ***Euonymus***

Pronunciation: (yew-ON-i-mus)

Species Name (Latin): *alatus*

Common Name: **Burning Bush,** Winged Spindle Tree

Family Name (Latin and Common): *Celastraceae, the Staff-Tree family*

Related Family Members in Book: *Celastrus*

Availability: commercial product: limited

Color: grayish brown

Unique Characteristics: Dried material has 4-angled branch with corky wings.

Design Applications: (medium) (line) Branch material provides an interesting line. Use in supporting armatures for European designs.

Longevity: long lasting

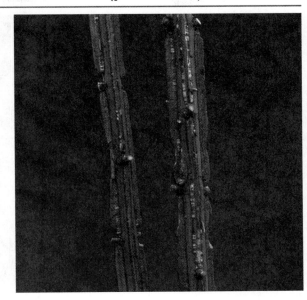

Genus Name (Latin): ***Euonymus***

Pronunciation: (yew-ON-i-mus)

Species Name (Latin): ***fortunei* 'Ivory Jade'**

Common Name: **Wintercreeper Euonymus**

Family Name (Latin and Common): *Celastraceae, the Staff-Tree family*

Related Family Members in Book: *Celastrus*

Availability: landscape product: availability varies

Color: green with creamy white margins

Unique Characteristics: Stem is curved with densely grouped leaves; other *Euonymus* available commercially.

Design Applications: (small) (filler) Nice choice for filler. Variegated leaves add interest.

Longevity: weeks 2-3 long lasting

Genus Name (Latin): ***Euphorbia***

Pronunciation: (yew-FOR-bee-a)

Species Name (Latin): ***marginata***

Common Name: **Snow-on-the-Mountain,** Ghostweed

Family Name (Latin and Common): *Euphorbiaceae,* the Spurge family

Related Family Members in Book: *Acalypha, Codiaeum,* and *Sapium*

Availability: commercial product: year-round

Color: green with white margins

Unique Characteristics: Stems release white milky sap when cut, sometimes causing severe burns or dermatitis in some individuals; seal stems to increase vase life.

Design Applications: (medium) (filler) Stems provide a nice accent and brighten focal zones. Interesting in combination with other mixed foliages.

Longevity: days 5-7

Genus Name (Latin): ***Fagus***

Pronunciation: (FAY-gus)

Species Name (Latin): ***sylvatica***

Common Name: **European Beech**

Family Name (Latin and Common): *Fagaceae,* the Beech family

Related Family Members in Book: *Quercus*

Availability: commercial product: limited May-October

Color: green

Unique Characteristics: Stem is woody; leaf is naturally glossy.

Design Applications: (large) (mass) Branches are useful for backgrounds. Longer laterals and individual leaves can work well in bridal designs or smaller-scale arrangements.

Longevity: days 7-10

Genus Name (Latin): *Fagus*   Pronunciation: (FAY-gus)

Species Name (Latin): *sylvatica*

Common Name: **European Beech**

Family Name (Latin and Common): *Fagaceae,*
the Beech family

Related Family Members in Book: *Quercus*

Availability: commercial product: varies
August-November

Color: orange, burgundy, rust, teal, and wine

Unique Characteristics: Foliage has woody stem;
fresh-cut foliage is dyed in several colors;
prolonged exposure to moisture causes dye to
exude from stem and leaf.

Design Applications: (large) (mass) Branches
provide mass. Leaves are nice when combined
with other dried or fresh products in traditional design styles.

Longevity: long lasting

Genus Name (Latin): *Ficus*   Pronunciation: (FY-kus)

Species Name (Latin): *benjamina*

Common Name: *Ficus,* Weeping Fig, Laurel,
Weeping Laurel

Family Name (Latin and Common): *Moraceae,*
the Mulberry family

Related Family Members in Book: none

Availability: indoor plant: year-round

Color: green (dark)

Unique Characteristics: Leaves glossy green;
commercially available as a greenhouse plant;
sap may cause skin to break out in some
people.

Design Applications: (small) (filler) Ovate-
elliptic leaf form works well in corsage and
boutonniere work. Plant is good for temporary
displays, dish gardens, and interior
plantscapes.

Longevity: days 3-5 as a cut foliage

Genus Name (Latin): *Ficus*

Pronunciation: (FY-kus)

Species Name (Latin): *benjamina 'Variegata'*

Common Name: **Variegated Ficus,** Variegated Weeping Fig

Family Name (Latin and Common): *Moraceae,* the Mulberry family

Related Family Members in Book: none

Availability: indoor plant: year-round

Color: green with white variegation

Unique Characteristics: Glossy leaf; commercially available as a greenhouse plant; sap may cause skin to break out in some people.

Design Applications: (small) (filler) Leaf form is unique for a filler. Use in corsage and boutonniere work. Plant makes a good choice for temporary landscape displays and dish garden combinations.

Longevity: days 3-5, as a cut foliage

Genus Name (Latin): *Ficus*

Pronunciation: (FY-kus)

Species Name (Latin): *lyrata*

Common Name: **Fiddle-Leaf Fig,** Fiddle-Leaf

Family Name (Latin and Common): *Moraceae,* the Mulberry family

Related Family Members in Book: none

Availability: indoor plant: year-round

Color: green (medium)

Unique Characteristics: Leaves are fiddle-shaped; commercially available as a large greenhouse plant.

Design Applications: (large) (mass) Leaves add visual mass. Big plant is useful for temporary landscape displays in combination with others having varying leaf forms.

Longevity: varies as a cut foliage

Genus Name (Latin): ***Galax***   Pronunciation: (GAY-laks)

Species Name (Latin): *urceolata*

Common Name: ***Galax***

Family Name (Latin and Common): *Diapensi-aceae,* the Diapensia family

Related Family Members in Book: none

Availability: commercial product: year-round with varying supplies in May-June

Color: green (medium), sometimes bronzing in fall or winter months

Unique Characteristics: Leaves are rounded or heart shaped on slender stem; can be cut into shapes or used in mass to produce a rosette.

Design Applications: (small-medium) (accent) Leaf is a staple for covering mechanics. Useful for textural interest with other shapes of foliage.

Longevity: weeks 1-2

Genus Name (Latin): ***Gaultheria***   Pronunciation: (gawl-THEE-ree-a)

Species Name (Latin): *shallon*

Common Name: **Salal,** Lemonleaf, Shallon

Family Name (Latin and Common): *Ericaceae,* the Heath family

Related Family Members in Book: *Erica, Rhodo-dendron,* and *Vaccinium*

Availability: commercial product: year-round

Color: green (medium)

Unique Characteristics: Air-dries well, becoming lighter gray-green in color.

Design Applications: (medium) (mass) Durable leaf is a staple for the florist. This foliage can be cut, glued, pressed, or dried. Full stem or pieces can be used as background, mass, or filler to both traditional and contemporary designs.

Longevity: weeks 2-3 long lasting

Genus Name (Latin): *Ginkgo*   Pronunciation: (GINGK-go)

Species Name (Latin): *biloba*

Common Name: *Ginkgo,* Maidenhair Tree

Family Name (Latin and Common): *Ginkgoaceae,* the Ginkgo family

Related Family Members in Book: none

Availability: landscape plant: varies

Color: green (medium), golden yellow in fall

Unique Characteristics: Foliage is distinctive, and is a personal favorite for its history; seed kernels, called ginkgo nuts, are widely eaten in the Orient, even though seed oil sometimes causes dermatitis. Mature, fallen fruits of the female tree produce offensive odor (butyric acid or rancid butter); typically not commercially available as a cut foliage; leaves are excellent for pressing.

Design Applications: (small) (form) Two-lobed leaf is a nice accent for pressed flower design. Also useful as a covering technique when glued to flat surfaces, creating a fabric appearance.

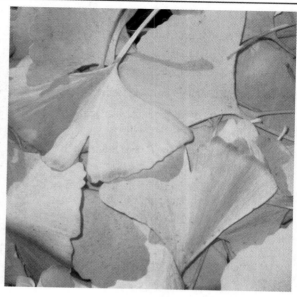

Longevity: varies

Genus Name (Latin): *Grevillea*   Pronunciation: (gre-VIL-ee-a)

Species Name (Latin): spp.

Common Name: *Grevillea*

Family Name (Latin and Common): *Proteaceae,* the Protea family

Related Family Members in Book: *Banksia, Leucadendron, Leucospermum, Protea, Serruria,* and *Telopea*

Availability: commercial product: varies September-November

Color: orange, yellow, green, and red

Unique Characteristics: Cut foliage form displays slender, serrated leaf margins; air-dried leaves continue to curl, adding to its unique appearance; available stem-dyed.

Design Applications: (small) (filler) Slender leaf has exotic appearance, providing excellent texture to fall and winter designs. Sometimes can act as a line element.

Longevity: weeks 1-2

Genus Name (Latin): ***Hedera***   Pronunciation: (HED-ur-a)

Species Name (Latin): ***helix*** cultivar

Common Name: **English Ivy**

Family Name (Latin and Common): *Araliacee,* the Aralia or Ginseng family

Related Family Members in Book: none

Availability: commercial product: year-round

Color: green

Unique Characteristics: Stem is vining with multilobed leaves; several cultivars available year-round as a greenhouse crop.

Design Applications: (medium) (form) Vines are used for adding length to bouquets, container arrangements, and garlands. Lobed leaf is a nice form for boutonniere, corsage, or headpiece designs. Plant is extensively used in container gardens.

Longevity: days 3-4, as a cut foliage

Genus Name (Latin): ***Hedera***   Pronunciation: (HED-ur-a)

Species Name (Latin): ***helix*** 'Variegata'

Common Name: **Variegated Ivy**

Family Name (Latin and Common): *Araliaceae,* the Aralia or Ginseng family

Related Family Member in Book: none

Availability: commercial product: year-round

Color: green with white margins

Unique Characteristics: Stem has graceful vining habit with multilobed leaves; also available as a greenhouse crop.

Design Applications: (medium) (form) Vine is useful for adding interesting line to cascades and other handheld bouquets. Effectively adds length to container arrangements and garlands. Variegated leaf is excellent for use in corsages, boutonnieres, and headpieces.

Longevity: days 3-5

Genus Name (Latin): ***Hedera***   Pronunciation: (HED-ur-a)

Species Name (Latin): spp.

Common Name: **Tree Ivy**

Family Name (Latin and Common): *Araliaceae,* the Aralia or Ginseng family

Related Family Members in Book: none

Availability: commercial product: year-round

Color: green (medium) with white margins

Unique Characteristics: Stem is woody, with ovate leaves of various sizes.

Design Applications: (large) (filler) Variegated foliage is a good filler. Useful for adding interest or lightening up focal zones. Pretty when combined with other foliages in all-green colored bouquets.

Longevity: days 7-10

Genus Name (Latin): ***Heuchera***   Pronunciation: (HEW-kur-a)

Species Name (Latin): *micrantha* **'Palace Purple'**

Common Name: **Palace Purple Heuchera**

Family Name (Latin and Common): *Saxifragaceae,* the Saxifrage family

Related Family Members in Book: *Astilbe, Hydrangea*

Availability: landscape plant: varies April-September

Color: purple (reddish), beet-red beneath

Unique Characteristics: Leaf is crinkled and deeply colored; flowers from several species are commercially available, but this plant is also quite striking for containerized gardens.

Design Applications: (medium) (accent) Leaf adds an interesting form and color to temporary landscapes or containerized gardens.

Longevity: varies

Genus Name (Latin): ***Hosta***   Pronunciation: (HOS-ta)

Species Name (Latin): *lancifolia*

Common Name: ***Hosta,*** Plantain Lily

Family Name (Latin and Common): *Liliaceae,* the Lily family

Related Family Members in Book: *Asparagus, Aspidistra, Convallaria, Gloriosa, Hemerocallis, Hyacinthus, Kniphofia, Liriope, Muscari Ornithogalum, Ruscus, Sandersonia, Scilla, Smilax, Tulipa,* and *Xerophyllum*

Availability: landscape plant: limited commercial availability

Color: green (dark)

Unique Characteristics: Foliage is narrowly lance shaped and glossy with thin stem; typically available as a landscape plant but periodically pops up in the commercial trade as a cut foliage.

Design Applications: (small) (mass) Leaf is a different choice for adding mass.

Longevity: varies

Genus Name (Latin): ***Hosta***   Pronunciation: (HOS-ta)

Species Name (Latin): **'Love Pat'**

Common Name: ***Hosta,*** Plantain Lily

Family Name (Latin and Common): *Liliaceae,* the Lily family

Related Family Members in Book: *Asparagus, Aspidistra, Convallaria, Gloriosa, Hemerocallis, Hyacinthus, Kniphofia, Liriope, Muscari, Ornithogalum, Ruscus, Sandersonia, Scilla, Smilax, Tulipa,* and *Xerophyllum*

Availability: landscape plant: commercially limited during summer months

Color: blue-green

Unique Characteristics: Foliage is heart shaped, cupped, and strongly puckered; typically available as a landscape plant but periodically appears as a commercial product in varying markets.

Longevity: varies

Design Applications: (large) (mass) Stiff leaf provides interesting color, working nicely with larger pink or white garden flowers.

Genus Name (Latin): ***Hosta*** | Pronunciation: (HOS-ta)

Species Name (Latin): **'Ginko Craig'**

Common Name: ***Hosta,*** Plantain Lily

Family Name (Latin and Common): *Liliaceae,* the Lily family

Related Family Members in Book: *Asparagus, Aspidistra, Convallaria, Gloriosa, Hemerocallis, Hyacinthus, Kniphofia, Liriope, Muscari, Ornithogalum, Ruscus, Sandersonia, Scilla, Smilax, Tulipa,* and *Xerophylllum*

Availability: landscape plant: limited commercial availability

Color: green (dark) with margins clear-white

Unique Characteristics: Foliage is lance shaped; typically available as a landscape plant but periodically appears as a cut foliage; many cultivars of Hosta are available with varying shades of color.

Design Applications: (small) (mass) Variegated leaf is nice for adding mass and accent.

Longevity: varies

Genus Name (Latin): ***Hosta*** | Pronunciation: (HOS-ta)

Species Name (Latin): **'Gold Standard'**

Common Name: ***Hosta,*** Plantain Lily

Family Name (Latin and Common): *Liliaceae,* the Lily family

Related Family Members in Book: *Asparagus, Aspidistra, Convallaria, Gloriosa, Hemerocallis, Hyacinthus, Kniphofia, Liriope, Muscari, Ornithogalum, Ruscus, Sandersonia, Scilla, Smilax, Tulipa,* and *Xerophyllum*

Availability: landscape plant: limited commercial availability

Color: yellow-green

Unique Characteristics: Foliage is uniquely yellow, ovate to heart shaped; typically available as a landscape plant but periodically appears as cut foliage; many varieties of Hosta are available with varying colors of green, yellow-green, or blue-green hues.

Longevity: varies

Design Applications: (medium) (mass) Leaf is a nice addition to designs of all green colored foliage.

Genus Name (Latin): **Hosta**

Pronunciation: (HOS-ta)

Species Name (Latin): spp.

Common Name: **Hosta,** Plantain Lily

Family Name (Latin and Common): *Liliaceae,* the Lily family

Related Family Members in Book: *Asparagus, Aspidistra, Convallaria, Gloriosa, Hemerocallis, Hyacinthus, Kniphofia, Liriope, Muscari, Ornithogalum, Ruscus, Sandersonia, Scilla, Smilax, Tulipa,* and *Xerophyllum*

Availability: landscape plant: limited commercial availability

Color: white with green margins

Unique Characteristics: Foliage has centrally marked white leaf with wide green margins; typically available as a landscape plant but periodically appears as a commercial cut foliage. This is one of five varieties in the book, which highlights their varying colorations and forms.

Longevity: varies

Design Applications: (medium-large) (mass) Leaf is a welcome addition to garden-style designs. Useful for adding interest and mass.

Genus Name (Latin): **Humulus**

Pronunciation: (HEW-mew-lus)

Species Name (Latin): *lupulus*

Common Name: **Hops**

Family Name (Latin and Common): *Cannabidaceae,* the Hemp family

Related Family Members in Book: none

Availability: commercial product: year-round

Color: green, red, yellow, and purple

Unique Characteristics: Dried material; holds conelike flower; appears commercially in preserved and stem-dyed garlands; has a slight odor that some find offensive. Also available as fresh-cut vine as seen in photo.

Design Applications: (large) (texture) Leaf and unique fruit are an interesting material for decorating. Single stems have a unique habit that works well for arbors, trellises, and handmade armatures. Garland is beautiful for home décor.

Longevity: days as a cut foliage; long lasting as a preserved material

Genus Name (Latin): **Hypoestes**   Pronunciation: (hy-po-EE-stes)

Species Name (Latin): *phyllostachya* 'Pink Splash'

Common Name: **Polka-Dot Plant,** Pink Polka-Dot Plant

Family Name (Latin and Common): *Acanthaceae,* the Acanthus family

Related Family Members in Book: *Acanthus, Pachysandra, Thungergia*

Availability: indoor plant: year-round

Color: green with pink dots

Unique Characteristics: Plant is available as an annual bedding plant but is often found as a greenhouse plant throughout the year; several varieties available; appearance is distinctive.

Design Applications: (small) (accent) Unusual leafed plant is useful in containerized gardens, including window box displays and dish gardens as an accent plant.

Longevity: varies

Genus Name (Latin): **Hypogymnia**   Pronunciation: (hy-po-GYM-nee-a)

Species Name (Latin): sp.

Common Name: **Lichen**

Family Name (Latin and Common): *Parmeliaceae,* the Skull Lichen family

Related Family Members in Book: *Letharia, Usnea*

Availability: commercial product: year-round

Color: gray with black underneath

Unique Characteristics: This dried material is one of several lichen species with similar characteristics.

Design Applications: (small) (texture) Dried material has unique characteristics, and is an excellent choice for adding textural interest to basal zones. Also useful as a covering material.

Longevity: long lasting

Genus Name (Latin): *Ilex*   Pronunciation: (EYE-leks)

Species Name (Latin): *opaca*

Common Name: **American Holly**

Family Name (Latin and Common): *Aquifoliaceae,* the Holly family

Related Family Members in Book: none

Availability: commercial product: November-December

Color: green (dark)

Unique Characteristics: Form of glossy leaf is distinctive with spiky margins; often available with red berries.

Design Applications: (small) (texture) Stiff leaf on woody stem makes this foliage an excellent textural addition to mixed seasonal conifer arrangements. Individual leaves are used for holiday body flower designs after removing spikes.

Longevity: weeks 1-3 long lasting

Genus Name (Latin): *Ilex*   Pronunciation: (EYE-leks)

Species Name (Latin): *aquifolium*

Common Name: **Variegated Holly**, English Holly

Family Name (Latin and Common): *Aquifoliaceae,* the Holly family

Related Family Members in Book: none

Availability: commercial product: November-December

Color: green with white margins

Unique Characteristics: Form is distinctive with spiky leaf margins; often available with red berries.

Design Applications: (small) (texture) Variegated leaf produces an excellent textural contrast in combination with other cut conifers. Individual leaves are used for holiday body flower designs after removing spikes.

Longevity: weeks 1-3 long lasting

Genus Name (Latin): ***Juniperus***   Pronunciation: (joo-NIP-er-us)

Species Name (Latin): ***communis***

Common Name: **Juniper,** Common Juniper

Family Name (Latin and Common): *Cupressaceae,* the Cypress family

Related Family Members in Book: *Calocedrus, Chamaecyparis,* and *Thuja*

Availability: commercial product: November-January

Color: bluish-green, often with berries

Unique Characteristics: Berries are bluish-gray; needlelike or scalelike leaves are prickly to touch, sometimes causing people skin irritation.

Design Applications: (medium) (texture) Branching material often contains berries, providing accent and textural interest.

Longevity: weeks 3-4 long lasting

Genus Name (Latin): ***Larix***   Pronunciation: (LAIR-iks; LAY-riks)

Species Name (Latin): ***occidentalis***

Common Name: **Western Larch**

Family Name (Latin): *Pinaceae,* the Pine family

Related Family Members in Book: *Abies, Picea, Pinus, Pseudotsuga* and *Tsuga*

Availability: landscape plant: limited as a commercial cut foliage

Color: green—light green in spring, dark green in summer, ochre-yellow in fall

Unique Characteristics: Needle is medium-fine; deciduous conifer has drooping branchlets.

Design Applications: (medium) (accent) Long, drooping branchlets have a wonderfully soft appearance. Excellent choice for contemporary designs to highlight unique form.

Longevity: days 5-7 as a cut foliage

Genus Name (Latin): ***Laurus*** | Pronunciation: (LAW-rus)

Species Name (Latin): sp.

Common Name: **Laurel**, Sweet Bay, Bay

Family Name (Latin and Common): *Lauracee,* the Laurel family

Related Family Members in Book: none

Availability: commercial product: varies year-round

Color: green (medium)

Unique Characteristics: Fragrant; leaves are used in cooking and yield an essential oil that is used in perfumery and medicine; can be air-dried.

Design Applications: (small) (filler) This filler is useful in everyday designs. Best when used in small amounts to accent its character. Nice for a fragrant addition to dried kitchen wreaths.

Longevity: days 7-10

Genus Name (Latin): ***Letharia*** | Pronunciation: (le-THAY-ree-a)

Species Name (Latin): ***vulpina***

Common Name: **Chartreuse Lichen**

Family Name (Latin and Common): *Parmeliaceae,* the Skull Lichen family

Related Family Members in Book: *Hypogymnia, Usnea*

Availability: commercial product: year-round

Color: green (chartreuse)

Unique Characteristics: Dried material; has intense color.

Design Applications: (small) (accent) Dried material's unique color draws attention. Makes an excellent choice for adding accent, and often is used for adding texture to basal focal zones in container arrangements.

Longevity: long lasting

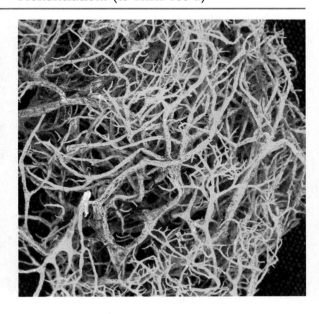

Genus Name (Latin): **Leucadendron**　　Pronunciation: (loo-ka-DEN-dron)

Species Name (Latin): *argenteum*

Common Name: **Silver Tree**

Family Name (Latin and Common): *Proteaceae,* the Protea family

Related Family Members in Book: *Banksia, Leucospermum, Protea, Serruria,* and *Telopea*

Availability: commercial product: varies year-round

Color: gray (silvery)

Unique Characteristics: Stem of leaves is densely covered with silvery, silky pubescence.

Design Applications: (large) (accent) Stem of distinctive silvery leaves can sometimes be used as a line element. Always provides accent because of its unique characteristics. Leaves can be removed from stem and used as a covering material for mechanics of various forms. Best to be used in contemporary design styles.

Longevity: weeks 2-3 long lasting

Genus Name (Latin): **Liriope**　　Pronunciation: (li-RY-oh-pee)

Species Name (Latin): *muscari*

Common Name: **Florists' Monkey Grass,** Big Blue Lilyturf

Family Name (Latin and Common): *Liliaceae,* the Lily family

Related Family Members in Book: *Asparagus, Aspidistra, Convallaria, Gloriosa, Hemerocallis, Hyacinthus, Kniphofia, Ornithogalum, Ruscus, Sandersonia, Smilax, Tulipa,* and *Xerophyllum*

Availability: commercial product: year-round

Color: green (dark)

Unique Characteristics: Leaves are wider than *L. spicata.*

Design Applications: (small to medium) (accent) Long, slender leaf works well in contemporary bridal bouquets. Excellent specimen to create sheltering effects in designs. Can also be used to add physical or visual movement.

Longevity: days 7-10

Genus Name (Latin): ***Liriope***   Pronunciation: (li-RY-oh-pee)

Species Name (Latin): **spicata** syn. *Ophiopogon spicatus*

Common Name: ***Liriope,*** Lilyturf, Creeping Lilyturf

Family Name (Latin and Common): *Liliaceae,* the Lily family

Related Family Members in Book: *Asparagus, Aspidistra, Convallaria, Gloriosa, Hemerocallis, Hyacinthus, Kniphofia, Ornithogalum, Ruscus, Sandersonia, Smilax, Tulipa,* and *Xerophyllum*

Availability: landscape plant: limited as a commercial cut foliage

Color: green (dark)

Unique Characteristics: Leaves are much smaller and narrower than *L. muscari.*

Design Applications: (small) (accent) Small, narrow blades can be useful in bridal designs and corsage work.

Longevity: days 5-7 as a cut foliage

Genus Name (Latin): ***Lobaria***   Pronunciation: (lo-BAY-ree-a)

Species Name (Latin): sp.

Common Name: **Foliose**

Family Name (Latin and Common): *Lobariaceae,* the Tree Lungwort family

Related Family Members in Book: none

Availability: commercial product: year-round

Color: brown, varying light to medium

Unique Characteristics: Dried material; if moistened it will regain more color and become pliable; appearance is undeniable.

Design Applications: (small) (texture) Pieces of this material add wonderful texture to basal focal zones of container arrangements, and is a good choice for adding accent. Also used to cover mechanics.

Longevity: long lasting

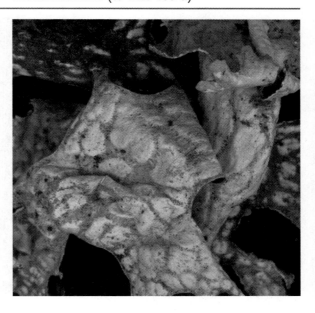

Genus Name (Latin): *Lonicera*   Pronunciation: (lo-NIS-ur-a)

Species Name (Latin): spp.

Common Name: **Honeysuckle Vine**

Family Name (Latin and Common): *Caprifoliaceae,* the Honeysuckle family

Related Family Members in Book: *Kolkwitzia, Symphoricarpos, Viburnum,* and *Weigela*

Availability: commercial product: year-round

Color: brown, variations of light brown and tan

Unique Characteristics: Dried material (vine) produces nice wreath forms; photo shows commercially available bundle.

Design Applications: (large) (texture) Groupings of thin vine can be shaped into varying sizes of wreaths. Long pieces when added to fresh or dried designs can create visual movement or rhythm.

Longevity: long lasting

Genus Name (Latin): *Lophomyrtus*   Pronunciation: (lo-fo-MUR-tus)

Species Name (Latin): *bullata*

Common Name: *Lophomyrtus*

Family Name (Latin and Common): *Myrtaceae,* the Myrtle family

Related Family Members in Book: *Callistemon, Chamelaucium, Eucalyptus, Leptospermum, Melaleuca, Myrtus,* and *Thryptomene*

Availability: commercial product: December-March

Color: red (bronze to red tinted)

Unique Characteristics: Fragrance is slight, sometimes offensive; ovate or rounded leaves are strongly puckered between veins.

Design Applications: (small) (line) Leaves have distinctive color and form. Linear material is useful for contemporary and traditional design styles. Makes a nice accent for bridal bouquets, body flowers, and everyday arrangements.

Longevity: days 6-10

Genus Name (Latin): ***Lunaria***   Pronunciation: (loo-NAY-ree-a)

Species Name (Latin): ***annua***

Common Name: **Money Plant,** Honesty, Silver-Dollars

Family Name (Latin and Common): *Brassicaceae,* the Mustard family

Related Family Members in Book: *Alysum, Brassica, Iberis, Matthiola*

Availability: commercial product: year-round

Color: white

Unique Characteristics: Dried material; round "silver dollars" shown, have appeared after the valves and seeds drop late in the season; very fragile.

Design Applications: (medium) (accent) Stem of papery texture adds interest to mixed dried flowers. Also works well in mass alone.

Longevity: long lasting

Genus Name (Latin): ***Lycopodium***   Pronunciation: (ly-ko-PO-dee-um)

Species Name (Latin): sp.

Common Name: ***Lycopodium,*** Club Moss

Family Name (Latin and Common): *Lycopodiaceae,* the Lycopodium family

Related Family Members in Book: none

Availability: commercial product: year-round

Color: green (bright)

Unique Characteristics: Form is distinctive; scalelike needles appear on many forked or horizontal branches.

Design Applications: (medium) (line) Linear material is used extensively in contemporary tropical arrangements and in combination with other exotics.

Longevity: days 7-10

Genus Name (Latin): ***Magnolia***

Pronunciation: (mag-NO-lee-a)

Species Name (Latin): ***grandiflora***

Common Name: ***Magnolia***

Family Name (Latin and Common): *Magnoliaceae,* the Magnolia family

Related Family Members in Book: none

Availability: commercial product: October-March

Color: green

Unique Characteristics: Leaf is glossy, dark green, often with brown tones beneath; stem is woody; leaves are also found preserved with glycerine and dyed for use in permanent botanical designs.

Design Applications: (large) (mass) Leaves are excellent for providing mass and useful for background.

Longevity: days 6-10

Genus Name (Latin): ***Mahonia***

Pronunciation: (ma-HO-nee-a)

Species Name (Latin): ***repens***

Common Name: ***Mahonia,*** Oregon Grape Holly

Family Name (Latin and Common): *Berberbidacee,* the Barberry family

Related Family Members in Book: *Nandina*

Availability: commercial product: varies year-round

Color: green (dark)

Unique Characteristics: Leaf is spiky and glossy.

Design Applications: (small) (filler) Spiky leaf is a good choice for adding texture and for use as a filler.

Longevity: days 5-10

Genus Name (Latin): ***Maranta***   Pronunciation: (mah-RAN-ta)

Species Name (Latin): *leuconeurea*
'Erythroneura'

Common Name: **Prayer Plant**

Family Name (Latin and Common):
Marantaceae, the Maranta or Arrowroot family

Related Family Members in Book: *Calathea*

Availability: commercial product: year-round

Color: green with colored venation

Unique Characteristics: Leaf has distinctive
venation; several *Maranta* are available for use
as cut foliage.

Design Applications: (medium) (accent) Leaf
with variegated designs and interesting venation add mass and accent. Makes an excellent
choice for contemporary container and bridal
bouquet designs.

Longevity: days 3-5

Genus Name (Latin): ***Melaleuca***   Pronunciation: (mel-a-LOO-ka)

Species Name (Latin): *armillaris*

Common Name: **Honey Bracelet,** Honey Myrtle

Family Name (Latin and Common): *Myrtaceae,*
the Myrtle family

Related Family Members in Book: *Callistemon,
Chamelaucium, Eucalyptus, Leptospermum,
Lophomyrtus, Myrtus,* and *Thryptomene*

Availability: commercial product: year-round

Color: green (medium to light)

Unique Characteristics: Stem is long and thin
with tiny linear leaves, often with white flowers that resemble a bottlebrush.

Design Applications: (small to large) (line)
Stems are an excellent choice for adding line
to varying scaled designs; also use as a filler
with interesting texture, color, and periodic
flowers.

Longevity: days 7-10

Genus Name (Latin): *Melissa*

Pronunciation: (mee-LIS-a)

Species Name (Latin): *officinalis*

Common Name: **Lemon Balm,** Common Balm, Sweet Balm, Bee Balm

Family Name (Latin and Common): *Lamiaceae,* the Mint family

Related Family Members in Book: *Lavandula, Leonotis, Mentha, Molucella, Monarda, Ocimum, Origanum, Perovskia, Physostegia, Rosmarinus, Salvia, Solenostemon, Stachys,* and *Thymus*

Availability: landscape plant: commercially limited as a commercial cut foliage

Color: green

Unique Characteristics: Fragrant, lemon-scented leaves can be pressed; also useful in potpourri.

Design Applications: (small) (texture) Foliage is an excellent choice for scented or green-colored bouquets.

Longevity: days 3-5, preferring clear water to floral foam

Genus Name (Latin): *Miscanthus*

Pronunciation: (mis-KAN-thus)

Species Name (Latin): *sinensis* 'Gracillimus'

Common Name: **Maiden Grass**

Family Name (Latin and Common): *Poaceae,* the Grass family

Related Family Members in Book: *Avena, Chasmanthium, Eleusine, Pennisetum, Phalaris, Phragmites, Phyllostachys, Setaria, Sorghum,* and *Triticum*

Availability: commercial product: limited August-October

Color: green (gray-green)

Unique Characteristics: Leaves are long, thin, and fine textured.

Design Applications: (texture) Narrow-leafed grass is nice when combined with bold-colored summer bloomers or in autumn bouquets combined with other grasses and exotics.

Longevity: days 5-7

Genus Name (Latin): ***Miscanthus***   Pronunciation: (mis-KAN-thus)

Species Name (Latin): *sinensis* **'Variegatus'**

Common Name: **Variegated Maiden Grass**

Family Name (Latin and Common): *Poaceae,* the Grass family

Related Family Members in Book: *Avena, Chasmanthium, Eleusine, Pennisetum, Phalaris, Phragmites, Phyllostachys, Setaria, Sorghum,* and *Triticum*

Availability: commercial product: limited August-October

Color: green and white

Unique Characteristics: Leaf has creamy white, longitudinal bands

Design Applications: (medium-large) (texture) Long grass can provide visual and physical movement to designs. Works beautifully in autumnal colored bouquets.

Longevity: days 3-5

Genus Name (Latin): ***Miscanthus***   Pronunciation: (mis-KAN-thus)

Species Name (Latin): *sinensis* **'Zebrinus'**

Common Name: **Zebra Grass**

Family Name (Latin and Common): *Poaceae,* the Grass family

Related Family Members in Book: *Avena, Chasmanthium, Eleusine, Pennisetum, Phalaris, Phragmites, Phyllostachys, Setaria, Sorghum,* and *Triticum*

Availability: commercial product: limited August-October

Color: green and yellow

Unique Characteristics: Leaf has yellow horizontal bands.

Design Applications: (texture) Long grass blades provide unique texture in combination with flowers, especially in monochromatic yellow designs. A different choice for providing visual and physical movement.

Longevity: days 3-5

Genus Name (Latin): ***Monstera***   Pronunciation: (MON-stur-a; mon-STEE-ra)

Species Name (Latin): ***deliciosa***

Common Name: ***Monstera,*** Swiss Cheese Plant

Family Name (Latin and Common): *Araceae,* the Arum family

Related Family Members in Book: *Anthurium, Caladium, Philodendron,* and *Zantedeschia*

Availability: commercial product: year-round

Color: green (dark)

Unique Characteristics: Form of leaf is unique; often contains large holes giving the appearance of Swiss cheese in more mature leaves; available in varying sizes.

Design Applications: (large) (form, mass) Leaf size adds visual mass. An excellent choice for high-style designs, and works well with tropical flowers.

Longevity: weeks 2-4 long lasting

Genus Name (Latin): ***Murraya***   Pronunciation: (moo-RAY-a)

Species Name (Latin): ***paniculata***

Common Name: **Coffee Foliage**

Family Name (Latin and Common): *Rutaceae,* the Rue family

Related Family Members in Book: *Poncirus*

Availability: commercial product: year-round

Color: green (dark)

Unique Characteristics: Leaf is glossy.

Design Applications: (small) (filler) Small leafed foliage is an excellent filler for adding interest. Useful for all design styles.

Longevity: days 7-10

Genus Name (Latin): ***Myrtus***

Pronunciation: (MUR-tus)

Species Name (Latin): ***communis***

Common Name: **Common Myrtle**

Family Name (Latin and Common): *Myrtaceae,* the Myrtle family

Related Family Members in Book: *Callistemon, Chamelaucium, Eucalyptus, Leptospermum, Lophomyrtus, Melaleuca,* and *Thryptomene*

Availability: commercial product: October-March

Color: green (dark)

Unique Characteristics: Fragrant, slightly lemon, when stems are broken; often available glycerine treated and stem dyed in other colors for use in permanent botanical designs.

Design Applications: (large) (line) Leafy line material is an excellent choice for traditional line mass style designs. Long lateral stems are often suitable for smaller-scale arrangements.

Longevity: weeks 1-2

Genus Name (Latin): ***Nandina***

Pronunciation: (nan-DY-na)

Species Name (Latin): ***domestica***

Common Name: ***Nandina,*** Heavenly Bamboo

Family Name (Latin and Common): *Berberidaceae,* the Barberry family

Related Family Members in Book: *Mahonia*

Availability: indoor plant: varies year-round as a commercial cut foliage

Color: green (medium) often with bronzing

Unique Characteristics: Fruit, tiny clustered berry, as seen in photo, is periodically available; leaves display a seasonal bronzing of color.

Design Applications: (medium) (texture) Good choice for texture, with berries adding accent. Use sparingly within a single design.

Longevity: days 5-10

Genus Name (Latin): *Nelumbo*	Pronunciation: (ne-LUM-bo)

Species Name (Latin): spp.

Common Name: **Lotus Pod**

Family Name (Latin and Common):
Nymphaeaceae, the Water Lily family

Related Family Members in Book: *Nymphaea*

Availability: commercial product: year-round

Color: brown

Unique Characteristics: Dried material is also available in dyed or spray-tinted colors; occasionally appears as a fresh-cut product with nice spring green coloration.

Design Applications: (large) (mass) Round form is useful for adding mass to traditional and contemporary designs. Works well when used with pine cones and other pods in holiday centerpieces or wreaths. An excellent choice for terracing.

Longevity: long lasting

Genus Name (Latin): *Nephrolepis*	Pronunciation: (nee-FROL-ep-is; ne-FROL-ep-is)

Species Name (Latin): *cordifolia*

Common Name: **Sword Fern**

Family Name (Latin and Common):
Polypodiaceae, the Polypody family

Related Family Members in Book: *Adiantum, Platycerium, Polystichum, Pteris,* and *Rumohra*

Availability: commercial product: year-round

Color: green (light)

Unique Characteristics: Fronds are linear, and can be pressed.

Design Applications: (small-medium) (line) Fern is often used for contemporary style containers and hand-held bouquets. A useful filler or line element.

Longevity: days 4-5

Genus Name (Latin): *Nyssa*

Pronunciation: (NIS-a)

Species Name (Latin): *sylvatica*

Common Name: **Black Tupelo,** Black Gum, Sour Gum

Family Name (Latin and Common): *Nyssaceae,* the Nyssa, Tupelo, or Sour-Gum family

Related Family Members in Book: none

Availability: landscape plant: typically not available as a commercial cut foliage

Color: green (dark) in summer, turning yellow, to orange, to purple in the fall

Unique Characteristics: Tree produces beautiful foliage with color in fall; glossy leaf can add interesting texture.

Design Applications: (medium) (line) Stem of glossy foliage can be a useful line element. When colored, it is an unusual choice for autumnal arrangements.

Longevity: varies

Genus Name (Latin): *Ocimum*

Pronunciation: (OH-kee-mum)

Species Name (Latin): *basilicum* **'Queen of Siam'**

Common Name: **Basil,** Sweet Basil

Family Name (Latin and Common): *Lamiaceae,* the Mint family

Related Family Members in Book: *Lavandula, Leonotis, Melissa, Mentha, Molucella, Monarda, Origanum, Perovskia, Physostegia, Rosmarinus, Salvia, Solenostemon, Stachys,* and *Thymus*

Availability: landscape plant: varies seasonally

Color: green

Unique Characteristics: Plant is a sweet herb; recently is more popular in cut-flower designs; provides an excellent purple flower head; other *Ocimum* foliages are available.

Design Applications: (medium) (accent) Leaf has interesting character that provides texture. Use in herbal bouquets. Flower should be used for accent.

Longevity: days 3-4 as a cut foliage

Genus Name (Latin): *Origanum*

Species Name (Latin): *vulgare*

Common Name: **Oregano**

Family Name (Latin and Common): *Lamiaceae,* the Mint family

Related Family Members in Book: *Lavandula, Leonotis, Melissa, Mentha, Molucella, Monarda, Ocimum, Perovskia, Physostegia, Rosmarinus, Salvia, Solenostemon, Stachys,* and *Thymus*

Availability: landscape plant: varies seasonally

Color: green

Unique Characteristics: Plant is a sweet herb; like other herbs, it has recently become more popular for cut-flower designs; mostly available as a bedding plant, but other *Ocimum* periodically appear as a commercial-cut foliage.

Design Applications: (small) (texture) Leafed plant is useful in herbal bouquets, providing a unique texture.

Longevity: days 3-4 as a cut foliage

Genus Name (Latin): *Pachysandra*

Species Name (Latin): sp.

Common Name: **Pachysandra,** Spurge

Family Name (Latin and Common): *Buxaceae,* the Box family

Related Family Members in Book: *Buxus*

Availability: landscape plant: limited as a commercial cut foliage

Color: green

Unique Characteristics: Plant is generally available as a ground cover, but can work well in cut-flower designs.

Design Applications: (small) (texture) Stem of clustered leaves can add texture to vase or container arrangements. Useful filler for garden flower bouquets.

Longevity: days 4-6 as a cut foliage

Genus Name (Latin): ***Pandanus*** Pronunciation: (pan-DAY-nus)

Species Name (Latin): *odoratissimus, utilis;* others

Common Name: **Hala,** Screw Pine

Family Name (Latin and Common): *Pandanaceae,* the Screw-Pine family

Related Family Members in Book: none

Availability: commercial product: year-round

Color: green with yellow striping

Unique Characteristics: Form is bold with length 24-36 inches; leaf stem is stiff. Can be bent into abstract shapes.

Design Applications: (large) (line) Bold material is an excellent choice for dramatic lines in high-style arrangements. Works well with tropical or exotic flowers, and is a good candidate for creating geometric shapes.

Longevity: weeks 2-3 long lasting

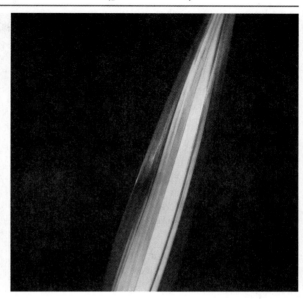

Genus Name (Latin): ***Papaver*** Pronunciation: (pa-PAY-ver)

Species Name (Latin): spp.

Common Name: **Poppy**

Family Name (Latin and Common): *Papaveraceae,* the Poppy family

Related Family Members in Book: none

Availability: commercial product: year-round

Color: brown

Unique Characteristics: Dried material is the seed head of blooming poppy; also is available as a fresh product in wonderful spring green color.

Design Applications: (small) (accent) Round form of interesting seed head combines texture and mass for adding accent. Use in smaller-scaled designs. Works well when combined with an assortment of pods and cones in everlasting designs.

Longevity: long lasting

Genus Name (Latin): ***Pelargonium***   Pronunciation: (pel-ar-GO-nee-um)

Species Name (Latin): spp.

Common Name: **Scented Geranium**

Family Name (Latin): *Geraniaceae,* the Geranium family

Related Family Members in Book: none

Availability: landscape plant: varies as a commercial cut foliage

Color: green

Unique Characteristics: Plant is fragrant; other *Pelargonium* are available with varying scents and leaf forms; some are very pubescent; used for annual bedding, and often found as a greenhouse crop.

Design Applications: (medium) (accent) Leaves make an interesting addition to bridal bouquets or small-scaled vase arrangements.

Longevity: days 2-4 as a cut foliage

Genus Name (Latin): ***Pennisetum***   Pronunciation: (pen-i-SEE-tum)

Species Name (Latin): ***setaceum* 'Purpureum'**

Common Name: **Purple Fountain Grass**

Family Name (Latin and Common): *Poaceae,* the Grass family

Related Family Members in Book: *Avena, Chasmanthium, Eleusine, Miscanthus, Phalaris, Phragmites, Phyllostachys, Setaria, Sorghum,* and *Triticum*

Availability: landscape plant: varies as a commercial cut foliage

Color: burgundy-purple foliage with mauve-rose spikes

Unique Characteristics: Plants are typically available in spring for summer bedding; plumes appear in August; *P. Setaceum* is also readily used for its green form.

Design Applications: (small) (texture) Multilinear plant will add texture to planting beds and container gardens. Makes a nice choice bundled for adding movement to cut flower designs.

Longevity: varies as a cut foliage

Genus Name (Latin): ***Phalaris***   Pronunciation: (FAL-a-ris)

Species Name (Latin): sp.

Common Name: ***Phalaris***

Family Name (Latin and Common): *Poaceae,* the Grass family

Related Family Members in Book: *Avena, Chasmanthium, Eleusine, Miscanthus, Pennisetum, Phragmites, Phyllostachys, Setaria, Sorghum,* and *Triticum*

Availability: commercial product: year-round

Color: tan

Unique Characteristics: Dried material has conical, rounded seed head with hairy tips; available in many colors, often changing with decorator trends.

Design Applications: (small) (filler) Dried material is best used when bundled. Works well in mixed dried bouquets, always providing texture and mass.

Longevity: long lasting

Genus Name (Latin): ***Philodendron***   Pronunciation: (fil-oh-DEN-dron)

Species Name (Latin): spp.

Common Name: ***Philodendron***

Family Name (Latin and Common): *Araceae,* the Arum family

Related Family Members in Book: *Anthurium, Caladium, Monstera,* and *Zantedeschia*

Availability: commercial product: year-round

Color: green (dark)

Unique Characteristics: Form is distinctively multilobed with a glossy appearance.

Design Applications: (medium) (form, mass) Leaf is a personal favorite for adding form, as well as mass, to high-style designs. Size makes it perfect for contemporary wedding bouquets. Works best with tropical flowers, and is a good choice for basal focal zones.

Longevity: days 5-7

Genus Name (Latin): ***Phormium***　　Pronunciation: (FOR-mee-um)

Species Name (Latin): ***tenax*** cultivar

Common Name: **Flax**, New Zealand Flax

Family Name (Latin and Common): *Agavaceae,* the Agave family

Related Family Members in Book: *Dracaena, Polianthes,* and *Yucca*

Availability: commercial product: year-round

Color: red-purple

Unique Characteristics: Color is reddish-purple, but green cultivars are also available; stiff midrib of leaf can be bent for angular patterns.

Design Applications: (medium) (line) Long linear leaf works well in geometric designs. Tropical flowers combine well with the appearance and coloration.

Longevity: weeks 1-2

Genus Name (Latin): ***Phyllostachys***　　Pronunciation: (fil-o-STAY-kis; fil-o-STAK-is)

Species Name (Latin): spp.

Common Name: **Bamboo**

Family Name (Latin and Common): *Poaceae,* the Grass family

Related Family Members in Book: *Avena, Chasmanthium, Eleusine, Miscanthus, Pennisetum, Phalaris, Phragmites, Setaria, Sorghum,* and *Triticum*

Availability: landscape plant: typically not available as a commercial cut foliage

Color: green

Unique Characteristics: Form is distinctive; leaves are susceptible to curling quickly out of water.

Design Applications: (large) (texture) Branching stem of leaves is perfect for Oriental designs.

Longevity: varies

Genus Name (Latin): ***Phyllostachys***   Pronunciation: (fil-oh-STAY-kis; fil-oh-STAK-is)

Species Name (Latin): spp.

Common Name: **River Cane**

Family Name (Latin and Common): *Poaceae,* the Grass family

Related Family Members in Book: *Avena, Chasmanthium, Eleusine, Miscanthus, Pennisetum, Phalaris, Phragmites, Setaria, Sorghum,* and *Triticum*

Availability: commercial product: year-round

Color: yellow-tan

Unique Characteristics: Dried material; hollow stems can hold water between cells.

Design Applications: (medium-large) (line) Slender leafless stems work well in Oriental designs. Use for producing geometric shapes in abstract or contemporary styles.

Longevity: long lasting

Genus Name (Latin): ***Physalis***   Pronunciation: (FY-sa-lis; FIS-a-lis)

Species Name (Latin): ***alkekengi***

Common Name: **Japanese Lantern**, Chinese Lantern

Family Name (Latin and Common): *Solanaceae,* the Nightshade family

Related Family Members in Book: *Brugmansia, Capsicum*

Availability: commercial product: varies late summer into fall

Color: orange

Unique Characteristics: Fragile dried material has pendulous papery-looking "lanterns"; displays bold color.

Design Applications: (medium) (accent) Stem of orange lanterns is a nice accent. Design use is best when allowed enough space to accentuate their unique habit.

Longevity: long lasting

Genus Name (Latin): ***Picea*** | Pronunciation: (PY-see-a; PIS-ee-a)

Species Name (Latin): ***abies***

Common Name: **Norway Spruce Cone**

Family Name (Latin and Common): *Pinaceae,* the Pine family

Related Family Members in Book: *Abies, Larix, Pinus, Pseudotsuga,* and *Tsuga*

Availability: commercial product: year-round

Color: brown (light)

Unique Characteristics: Dried material contains cone with scales that are without undulations.

Design Applications: (large) (mass) Cone is a nice addition to holiday wreaths and centerpieces. Size and shape are good for adding mass to traditional or contemporary design styles.

Longevity: long lasting

Genus Name (Latin): ***Picea*** | Pronunciation: (PY-see-a; PIS-ee-a)

Species Name (Latin): ***glauca***

Common Name: **White Spruce Cones**

Family Name (Latin and Common): *Pinaceae,* the Pine family

Related Family Members in Book: *Abies, Larix, Pinus, Pseudotsuga,* and *Tsuga*

Availability: commercial product: year-round

Color: brown (light)

Unique Characteristics: Dried material is smaller than other cones that are typically available commercial products.

Design Applications: (small) (mass) Cone is an excellent choice for adding interest through texture and mass. Makes a beautiful addition to holiday wreaths and centerpieces.

Longevity: long lasting

Genus Name (Latin): ***Picea***   Pronunciation: (PY-see-a; PIS-ee-a)

Species Name (Latin): spp.

Common Name: **Spruce**

Family Name (Latin and Common): *Pinaceae,* the Pine family

Related Family Members in Book: *Abies, Larix, Pinus, Pseudotsuga,* and *Tsuga*

Availability: landscape plant: typically not available as a commercial cut foliage

Color: green

Unique Characteristics: Displays a pendulous habit; other *Picea* readily appear as commercial cut foliage.

Design Applications: (medium) (filler) Needles are short. All spruce have texture that works well in combination with other conifers. Useful for holiday decorations.

Longevity: weeks 1-2

Genus Name (Latin): ***Pinus***   Pronunciation: (PY-nus)

Species Name (Latin): ***nigra*** subsp. ***austriaca***

Common Name: **Austrian Pine Cone,** Austriaca Pine Cone

Family Name (Latin and Common): *Pinaceae,* the Pine family

Related Family Members in Book: *Abies, Larix, Picea, Pseudotsuga,* and *Tsuga*

Availability: commercial product: year-round

Color: brown (light)

Unique Characteristics: Dried material; cone is rounded in shape, with spiny tips; size can reach 3 inches.

Design Applications: (medium) (mass) Cone has form, nicely suited for holiday wreaths and arrangements. Excellent in combination with other cones of mixed conifers.

Longevity: long lasting

Genus Name (Latin): *Pinus*   Pronunciation: (PY-nus)

Species Name (Latin): *strobus*

Common Name: **White Pine,** Eastern White
Pine

Family Name (Latin and Common): *Pinaceae,*
the Pine family

Related Family Members in Book: *Abies, Picea,
Larix, Pseudotsuga,* and *Tsuga*

Availability: commercial product: November-
January

Color: green (medium)

Unique Characteristics: Leaf is needlelike, long,
and thin; branches are flexible; white pine rop-
ing is commonly used for holiday decorations.

Design Applications: (medium) (texture, filler)
Branches of flexible needles provide soft looking texture to mixed conifer designs. Use for
holiday wreaths, container arrangements,
bridal bouquets, corsages, and boutonnieres.

Longevity: weeks 2-3 long lasting

Genus Name (Latin): *Pinus*   Pronunciation: (PY-nus)

Species Name (Latin): *strobus*

Common Name: **Eastern White Pine**

Family Name (Latin and Common): *Pinaceae,*
the Pine family

Related Family Members in Book: *Abies, Larix,
Picea, Pseudotsuga,* and *Tsuga*

Availability: commercial product: year-round

Color: brown (light)

Unique Characteristics: Dried material; resinous
cylindrical cone is often curved with apex
pointed; length is 4-6 inches.

Design Applications: (medium) (texture) Cone
is a staple for holiday designs. Works well with
mixed cut conifers and flowers in container
arrangements. Also suitable for use with permanent holiday decorations of all kinds.

Longevity: long lasting

Genus Name (Latin): ***Pinus*** Pronunciation: (PY-nus)

Species Name (Latin): ***taeda***

Common Name: **Loblolly Pine Cone**

Family Name (Latin and Common): *Pinaceae,*
the Pine family

Related Family Members in Book: *Abies, Larix,
Picea, Pseudotsuga,* and *Tsuga*

Availability: commercial product: year-round

Color: brown (light)

Unique Characteristics: Dried material; cylindrical cone 3-5 inches long or more.

Design Applications: (large) (mass) Cone adds
emphasis to holiday cut flower designs. Works
beautifully when grouped by themselves in
mass for merchandising displays.

Longevity: long lasting

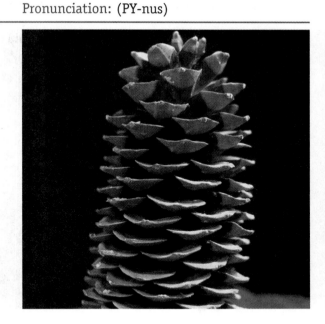

Genus Name (Latin): ***Pittosporum*** Pronunciation: (pi-TOS-po-rum; pit-o-SPO-rum)

Species Name (Latin): ***tenuifolium*** cultivar

Common Name: **Israeli Pitt**

Family Name (Latin and Common):
Pittosporaceae, the Pittosporum family

Related Family Members in Book: none

Availability: commercial product: year-round

Color: green with white margin

Unique Characteristics: Leaf margins are typically wavy and glossy; form differs from other
varieties in the book.

Design Applications: (small) (filler) Variegated
leaf works well in vase or container arrangements. Makes a nice choice for brightening
focal zones.

Longevity: days 7-10

Genus Name (Latin): ***Pittosporum***   Pronunciation: (pi-TOS-po-rum; pit-o-SPO-rum)

Species Name (Latin): ***tobira***

Common Name: ***Pittosporum,*** Japanese Mock Orange

Family Name (Latin and Common): *Pittosporaceae,* the Pittosporum family

Related Family Members in Book: none

Availability: commercial product: year-round

Color: green (medium)

Unique Characteristics: Leaf is obovate and semiglossy; durability is an asset.

Design Applications: (small) (filler) Round leaf form provides interest. Use as a filler and accent in container arrangements as well as wedding bouquets, corsages, and boutonnieres.

Longevity: weeks 1-2

Genus Name (Latin): ***Pittosporum***   Pronunciation: (pi-TOS-po-rum; pit-o-SPO-rum)

Species Name (Latin): ***tobira* 'Variegatum'**

Common Name: **Variegated Pittosporum,** Japanese Mock Orange

Family Name (Latin and Common): *Pittosporaceae,* the Pittosporum family

Related Family Members in Book: none

Availability: commercial product: year-round

Color: green with white margins

Unique Characteristics: Leaf is obovate and semiglossy; durability is an asset.

Design Applications: (small) (filler) Round leaf form provides filler and mass. Use in container arrangements as well as wedding bouquets, corsages, and boutonnieres. Variegated leaf can be used to lighten up dark focal zones.

Longevity: weeks 1-2

Genus Name (Latin): ***Platycerium***

Pronunciation: (plat-I-SEE-ree-um)

Species Name (Latin): spp.

Common Name: **Staghorn Fern**

Family Name (Latin and Common): *Polypodiaceae,* the Polypody family

Related Family Members in Book: *Adiantum, Nephrolepis, Polystichum, Pteris,* and *Rumohra*

Availability: commercial product: year-round

Color: green

Unique Characteristics: Leaf is forked into antlerlike lobes, covered with tiny starlike hairs; this epiphyte is also available as a greenhouse plant.

Design Applications: (large) (form) Lobed leaf is a personal favorite for accent. Unusual form should be used in high-style designs or in some manner that fully displays its distinctive form.

Longevity: days 3-5

Genus Name (Latin): ***Podocarpus***

Pronunciation: (po-do-KAR-pus; pod-o-KAR-pus)

Species Name (Latin): ***macrophyllus***

Common Name: ***Podocarpus,*** Southern Yew, Buddhist Pine

Family Name (Latin and Common): *Podocarpaceae,* the Podocarpus family

Related Family Members in Book: none

Availability: commercial product: year-round

Color: green (dark)

Unique Characteristics: Leaf is linear, lanceolate, and pointed on crowded twigs; foliage is durable; periodically is available with tiny blue berries.

Design Applications: (small) (filler) Linear leafed foliage is an interesting filler. Plays well against larger, rounder ovate leaf forms, such as *Gaultheria.*

Longevity: weeks 1-3 long lasting

Genus Name (Latin): ***Polygonatum***

Pronunciation: (po-lig-o-NAY-tum)

Species Name (Latin): *odoratum* **'Variegatum'**

Common Name: **Solomon's Seal**

Family Name (Latin and Common): *Liliaceae,* the Lily family

Related Family Members in Book: *Asparagus, Aspidistra, Convallaria, Gloriosa, Hemerocallis, Hosta, Hyacinthus, Kniphofia, Lillium, Liriope, Muscari, Ornithogalum, Ruscus, Sandersonia, Tulipa,* and *Xerophyllum*

Availability: landscape plant: varies commercially as a cut foliage in spring

Color: green with white margins

Unique Characteristics: Stem is arching with elliptic-ovate variegated leaves; small fragrant white flowers dangle beneath leaves in spring.

Design Applications: (medium) (filler) Arching habit of stem provides linear accent. Use in combination with other spring flowers, especially in vegetative systems of design.

Longevity: days 5-7

Genus Name (Latin): ***Polystichum***

Pronunciation: (po-LIS-ti-kum)

Species Name (Latin): *munitum*

Common Name: **Western Sword Fern,** Shield Fern, Flat Fern

Family Name (Latin and Common): *Polypodiaceae,* the Polypody family

Related Family Members in Book: *Adiantum, Nephrolepis, Platycerium, Pteris,* and *Rumohra*

Availability: commercial product: year-round with varying supply in summer months

Color: green (medium)

Unique Characteristics: Stems are 24-28 inches long; plant is similar to *Nephrolepis* but has longer and wider leaves, often containing spores beneath.

Design Applications: (large) (line) Foliage is a good choice for background or outline in line mass designs. Also useful as a filler.

Longevity: weeks 1-2

Genus Name (Latin): ***Poncirus*** syn. *Aegle*   Pronunciation: (pon-SY-rus)

Species Name (Latin): ***trifoliata***

Common Name: **Bell Cup**, Trifoliate Orange, Hardy Orange

Family Name (Latin and Common): *Rutaceae,* the Rue family

Related Family Members in Book: Murraya

Availability: commercial product: year-round

Color: brown

Unique Characteristics: Dried material is available in natural and many stained or dyed colors; also is available dried whole or cut and placed on wire stems as shown in photo.

Design Applications: (large) (mass) Large size adds visual weight to any arrangement.

Longevity: long lasting

Genus Name (Latin): ***Protea***   Pronunciation: (PRO-tee-ah)

Species Name (Latin): spp.

Common Name: ***Protea***

Family Name (Latin and Common): *Proteaceae,* the Protea family

Related Family Members in Book: *Banksia, Leucadendron, Leucospermum, Serruria,* and *Telopea*

Availability: commercial product: year-round

Color: brown (light)

Unique Characteristics: Dried material is available in many dyed or spray-tinted colors; shown on wired stem in photo.

Design Applications: (medium) (mass) The flat round form of this material is useful for providing mass in any design style. Works nicely when combined with cones and other dried pods in wreaths or holiday arrangements.

Longevity: long lasting

Genus Name (Latin): ***Protea***

Pronunciation: (PRO-tee-ah)

Species Name (Latin): spp.

Common Name: ***Protea***

Family Name (Latin and Common): *Proteaceae,* the Protea family

Related Family Members in Book: *Banksia, Leucadendron, Leucospermum, Serruria,* and *Telopea*

Availability: commercial product: year-round

Color: brown (light)

Unique Characteristics: Dried material is available in several dyed colors and on wired stem as shown in photo.

Design Applications: (large) (mass) Flat, round form adds mass to contemporary or traditional designs. Often is used in combination with permanent botanicals.

Longevity: long lasting

Genus Name (Latin): ***Protea***

Pronunciation: (PRO-tee-ah)

Species Name (Latin): ***repens***

Common Name: **Sugarbush Protea,** Honey Flower

Family Name (Latin and Common): *Proteaceae,* the Protea family

Related Family Members in Book: *Banksia, Leucadendron, Leucospermum, Serruria,* and *Telopea*

Availability: commercial product: year-round

Color: red

Unique Characteristics: Dried material; Outside surface is sometimes sticky.

Design Applications: (medium) (accent) Flower form and color are unique, adding accent and texture to container designs or wall decor. Natural red color is nice when combined with dried grasses or preserved conifers.

Longevity: long lasting

Genus Name (Latin): ***Pseudotsuga***   Pronunciation: (soo-doe-SOO-ga)

Species Name (Latin): ***menziesii***

Common Name: **Douglas Fir**

Family Name (Latin and Common): *Pinaceae,*
the Pine family

Related Family Members in Book: *Abies, Larix,
Picea, Pinus,* and *Tsuga*

Availability: commercial product: November-
January

Color: green (medium)

Unique Characteristics: Needles (leaves) are
flat, linear, having two white bands beneath.

Design Applications: (small) (filler) Needled
branches are nice when combined with several
other types of conifers. Use in container holiday arrangements of all styles.

Longevity: weeks 1-3 long lasting

Genus Name (Latin): ***Psylliostachys***   Pronunciation: (sy-lee-oh-STAK-us)

Species Name (Latin): ***suworowii***

Common Name: **Rattail Statice**

Family Name (Latin and Common):
Plumbaginaceae, the Plumbago or Leadwort
family

Related Family Members in Book: *Goniolimon,
Limonium*

Availability: commercial product: year-round

Color: pink

Unique Characteristics: Dried material has long,
slender stems; occasionally is available as a
commercial cut flower.

Design Applications: (small) (filler) Spike
flower is an unusual choice for accent and
filler. Works uniquely well in monochromatic
designs of everlasting flowers.

Longevity: long lasting

Genus Name (Latin): ***Pteris***   Pronunciation: (TEE-ris; TER-is)

Species Name (Latin): spp.

Common Name: **Brake Fern,** Table Fern

Family Name (Latin and Common):
Polypodiaceae, the Polypody family

Related Family Members in Book: *Adiantum,
Nephrolepis, Platycerium, Polystichum,* and
Rumohra

Availability: indoor plant: year-round

Color: green (medium)

Unique Characteristics: Leaf tips are ruffled;
leaf has an arching habit; makes an unusual
houseplant.

Design Applications: (small) (form) Leaf form
is unique. Choose this material for individual
fronds in simple budvases or contemporary
bridal designs. Plant is good for adding interest
to containerized garden mixtures.

Longevity: days 2-4 as a cut foliage

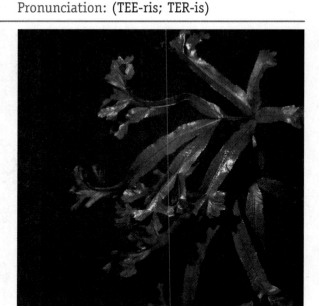

Genus Name (Latin): ***Punica***   Pronunciation: (PEW-ni-ka)

Species Name (Latin): ***granatum*** cultivar

Common Name: **Pomegranate**

Family Name (Latin and Common): *Puniaceae,*
the Pomegranate family

Related Family Members in Book: none

Availability: commercial product: year-round

Color: orange-red

Unique Characteristics: Dried material; typically available freeze-dried as shown; sometimes
also available in predried slices.

Design Applications: (medium) (mass) Round
form is a great choice for adding mass to
wreaths and permanent container arrangements. Makes a good choice for pillowing
technique. Looks wonderful with mixed fresh
or dried materials in autumnal color
combinations.

Longevity: long lasting

Genus Name (Latin): ***Quercus***   Pronunciation: (KWER-kus)

Species Name (Latin): *palustris*

Common Name: **Oak Leaves**

Family Name (Latin and Common): *Fagaceae,* the Beech family

Related Family Members in Book: *Fagus*

Availability: commercial product: year-round

Color: orange, yellow, rust, and green (lime)

Unique Characteristics: Dried material available has been stem-dyed and preserved with glycerine; avoid excessive moisture; other preserved *Quercus* are also available in colors.

Design Applications: (medium) (filler) Branch of medium-size pin-oak leaves are well suited for dried wreaths or arrangements. Most effective use is when leaves are gathered into small bunches. Recent introduction of bright green makes this product more adaptable for transseasonal uses.

Longevity: long lasting

Genus Name (Latin): ***Rhamnus***   Pronunciation: (RAM-nus)

Species Name (Latin): *alaternus* **'Argenteovariegata'**

Common Name: ***Rhamnus,*** Italian Buckthorn

Family Name (Latin and Common): *Rhamnaceae,* the Buckthorn family

Related Family Members in Book: none

Availability: commercial product: year-round

Color: green (gray-green)

Unique Characteristics: Leaf has leathery and glossy appearance with white margins.

Design Applications: (medium) (line) Small leaves on medium sized stem provides both accent and line. Use in traditional line or line-mass arrangements of all sizes.

Longevity: weeks 1-2

Genus Name (Latin): **Rhipsalis**   Pronunciation: (RIP-sa-lis)

Species Name (Latin): sp.

Common Name: **Rhipsalis**

Family Name (Latin and Common): *Cactaceae,* the Cactus family

Related Family Members in Book: none

Availability: indoor plant: year-round

Color: green (medium to light)

Unique Characteristics: Plant has succulent stems.

Design Applications: (medium) (accent) Appearance of stems always add interest. Works well in contemporary bridal and everyday designs.

Longevity: days 1-2

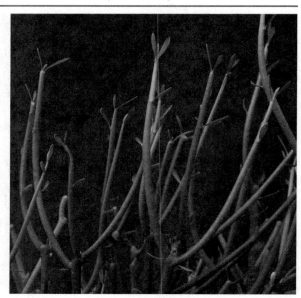

Genus Name (Latin): **Rhus**   Pronunciation: (RUS; ROOS)

Species Name (Latin): spp.

Common Name: **Sumac**

Family Name (Latin and Common): *Anacardiaceae,* the Cashew family

Related Family Members in Book: *Cotinus, Schinus*

Availability: landscape plant: commercially limited May-October as a commercial cut foliage

Color: green changing to orange, then purplish red in fall

Unique Characteristics: Colors are uniquely varied on one leaf; leaves are suitable for pressing; flower panicles are prominent as shown.

Design Applications: (large) (accent) Leaf form is distinctive. Always provides drama with its fall color.

Longevity: days 2-4 as a cut foliage

Genus Name (Latin): ***Rosmarinus***

Pronunciation: (ros-ma-RY-nus; roz-ma-RY-nus)

Species Name (Latin): ***officinalis*** cultivar

Common Name: **Rosemary**

Family Name (Latin and Common): *Lamiaceae,* the Mint family

Related Family Members in Book: *Lamium, Lavandula, Leonotis, Melissa, Mentha, Molucella, Monarda, Ocimum, Origanum, Physostegia, Salvia, Solenostemon, Stachys,* and *Thymus*

Availability: commercial product: summer months into fall as a cut foliage

Color: gray-green

Unique Characteristics: Fragrance is strong; fresh flower tops are used to distill the aromatic oil used in perfumery and medicine; dried leaves are used in seasoning. Plants are available year-round as a greenhouse crop, including some topiary shapes.

Design Applications: (small) (accent) Strong fragrance adds aroma to linear material. Stems

can be used in vegetative garden designs. Combines well with many other herbs.

Longevity: days 4-7

Genus Name (Latin): ***Rumohra***

Pronunciation: (roo-MO-rah)

Species Name (Latin): ***adiantiformis***

Common Name: **Leatherleaf,** Leatherleaf Fern, Baker Fern

Family Name (Latin and Common): *Polypodiaceae,* the Polypody family

Related Family Members in Book: *Adiantum, Nephrolepis, Platycerium, Polistichum,* and *Pteris*

Availability: commercial product: year-round

Color: green (medium to dark)

Unique Characteristics: Fronds are triangular; appearance is sometimes glossy; species is available in several grades of size.

Design Applications: (medium-large) (filler) Multipurpose foliage works well to cover mechanics, add mass, or texture.

Longevity: weeks 1-2 or more long lasting

Genus Name (Latin): *Ruscus*

Pronunciation: (RUS-kus)

Species Name (Latin): *hypoglossum*

Common Name: **Italian Ruscus,** Smilax Ruscus

Family Name (Latin and Common): *Liliaceae,* the Lily family

Related Family Members in Book: *Asparagus, Aspidistra, Convallaria, Gloriosa, Hemerocallis, Hosta, Hyacinthus, Kniphofia, Lilium, Liriope, Muscari, Sandersonia, Tulipa,* and *Xerophyllum*

Availability: commercial product: year-round

Color: green (dark)

Unique Characteristics: Leaf is elliptic to oblanceolate; durability is an asset.

Design Application: (medium) (line) Broad, dark green leaves create visual weight. Stem of leaves is slender enough to provide line to small arrangements.

Longevity: days 7-10

Genus Name (Latin): *Ruscus*

Pronunciation: (RUS-kus)

Species Name (Latin): *aculeatus*

Common Name: *Ruscus,* Holland Ruscus, Butcher's Broom

Family Name (Latin and Common): *Liliaceae,* the Lily family

Related Family Members in Book: *Asparagus, Aspidistra, Convallaria, Gloriosa, Hemerocallis, Hosta, Hyacinthus, Kniphofia, Lilium, Liriope, Muscari, Sandersonia, Tulipa,* and *Xerophyllym*

Availability: commercial product: year-round

Color: green (dark)

Unique Characteristics: Leaf is glossy, stiff, and spiny or pointed; stems are sometimes arching. This durable foliage is available in spray-tinted colors around the winter holidays.

Design Applications: (large) (line) Branched stem of glossy leaves makes this foliage an excellent choice for adding line and length. Lateral stems are useful in small-scaled container arrangements or body flower designs.

Longevity: days 7-10

Genus Name (Latin): *Sabal*

Pronunciation: (SAY-bal)

Species Name (Latin): spp.

Common Name: **Palmetto Palm**

Family Name (Latin and Common): *Arecaceae,* the Palm family

Related Family Members in Book: *Chamaedorea* and *Cocos*

Availability: commercial product: limited as a cut foliage

Color: green (medium)

Unique Characteristics: Palmate leaf form may be trimmed for interesting shapes; available as dried material already trimmed into fan design.

Design Applications: (large) (mass) Leaf adds mass. Use as background material in large displays.

Longevity: days 5-7

Genus Name (Latin): *Sabal*

Pronunciation: (SAY-bal)

Species Name (Latin): spp.

Common Name: *Sabal,* Palmetto

Family Name (Latin and Common): *Arecaceae,* the Palm family

Related Family Members in Book: *Chamaedorea* and *Cocos*

Availability: commercial product: year-round

Color: brown (natural)

Unique Characteristics: Dried material is available dyed or painted in decorator colors.

Design Applications: (large) (line) Interesting line material is nice for permanent botanical designs, especially when three or more are grouped together.

Longevity: long lasting

Genus Name (Latin): *Saccharum*

Pronunciation: (sak-KAR-um)

Species Name (Latin): *ravennae*

Common Name: **Plume Grass**

Family Name (Latin and Common): *Poaceae,* the Grass family

Related Family Members in Book: *Avena, Chasmanthium, Eleusine, Miscanthus, Pennisetum, Phalaris, Phragmites, Phyllostachys, Setaria, Sorghum,* and *Triticum*

Availability: commercial product: year-round

Color: tan

Unique Characteristics: Dried material consists of soft plume.

Design Applications: (large) (texture) Feathery appearance of the plume adds a delicate texture. Appearance is striking if grouped in mass.

Longevity: long lasting

Genus Name (Latin): *Salix*

Pronunciation: (SAY-liks)

Species Name (Latin): *discolor*

Common Name: **Pussy Willow**

Family Name (Latin and Common): *Salicaceae,* the Willow family

Related Family Members in Book: none

Availability: commercial product: November-May with varying supplies June-October

Color: brown

Unique Characteristics: Dried material is also available as a perishable product; dark stems with white fuzzy catkins are distinctive.

Design Applications: (small-large) (line) Stems are extremely versatile for adding line. Use in any design style. It is a favorite for spring arrangements and is often used to create framing or build armatures.

Longevity: long lasting

Genus Name (Latin): *Salix*   Pronunciation: (SAY-liks)

Species Name (Latin): *matsudana* 'Tortuosa'

Common Name: **Corkscrew Willow**, Dragon Claw Willow

Family Name (Latin and Common): *Salicaceae,* the Willow family

Related Family Members in Book: none

Availability: commercial product: year-round

Color: green (light)

Unique Characteristics: Branch is irregularly curved; stems occasionally will leaf out after prolonged periods in fresh water; also is available as a dried material.

Design Applications: (medium-large) (line) Branch material is an excellent choice for interesting line. Pliable stems can be bent to create tension or sheltering effects in designs. Appearance is very nice when used alone in mass, but is suitable for all styles of design from traditional line mass to high style.

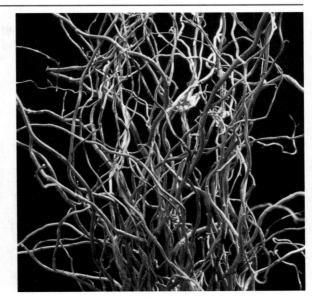

Longevity: weeks 3 or more long lasting

Genus Name (Latin): *Salix*   Pronunciation: (SAY-liks)

Species Name (Latin): spp.

Common Name: **Fasciated Willow**, Natraj

Family Name (Latin and Common): *Salicaceae,* the Willow family

Related Family Members in Book: none

Availability: commercial product: year-round

Color: brown (dark)

Unique Characteristics: Stem is contorted; product is very unusual with no two stems alike.

Design Applications: (large) (line) Thick stem is great for oriental or high style designs. Place where unique character can be observed.

Longevity: long lasting

Genus Name (Latin): **Salvia**   Pronunciation: (SAL-vee-a)

Species Name (Latin): *officinalis* 'Tricolor'

Common Name: **Tricolor Sage**

Family Name (Latin and Common): *Lamiaceae*, the Mint family

Related Family Members in Book: *Lavandula, Leonotis, Melissa, Mentha, Molucella, Monarda, Ocimum, Physostegia, Rosmarinus, Solenostemon, Stachys,* and *Thymus*

Availability: landscape plant: varies seasonally

Color: purple, green and white variegated

Unique Characteristics: Leaf is tricolored; available as a potted plant for indoor or outdoor use.

Design Applications: (small) (accent) Multicolored leaf works well with other herbs. Use plant to add accent in container gardens.

Longevity: days 3-4 as a cut foliage

Genus Name (Latin): **Sapium**   Pronunciation: (SAY-pee-um)

Species Name (Latin): *sebiferum*

Common Name: **Tallow Tree Berry,** Chinese Tallow Tree, Vegetable Tallow Tree

Family Name (Latin and Common): *Euphorbiaceae,* the Euphorbia family

Related Family Members in Book: *Acalypha, Codiaeum,* and *Euphorbia*

Availability: commercial product: varies year-round

Color: white

Unique Characteristics: Seeds are waxy.

Design Applications: (small) (filler) Clustered seed is a nice addition to both fresh or dried designs.

Longevity: long lasting

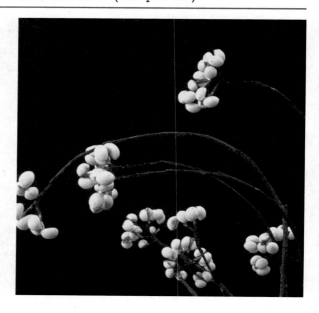

Genus Name (Latin): *Sarracenia*   Pronunciation: (sair-a-SEE-nee-a)

Species Name (Latin): spp.

Common Name: *Sarracenia,* Pitcher Plant

Family Name (Latin and Common): *Sarraceniaceae,* the Pitcher Plant family

Related Family Members in Book: none

Availability: commercial product: year-round

Color: white veined with green or purple

Unique Characteristics: Dried material has tubular-shaped leaves that are often mistaken for flowers; one side is winged and terminated by a lid; displays colored venation; several *Sarracenia* species are available.

Design Applications: (medium) (line) Material is wonderfully unique. Exquisite when grouped in mass allowing the unusual form to be highlighted, and a good candidate for geometric or parallel systems.

Longevity: days 7-10

Genus Name (Latin): *Schefflera*   Pronunciation: (shef-LEER-a)

Species Name (Latin): sp.

Common Name: *Schefflera,* Umbrella Tree

Family Name (Latin and Common): *Araliaceae,* the Ginseng family

Related Family Members in Book: *Hedera*

Availability: indoor plant: year-round

Color: green (dark)

Unique Characteristics: Leaf is palmate; cut foliage is durable.

Design Applications: (medium) (form) Leaf has different form for adding accent. Use in bridal bouquets or smaller container arrangements.

Longevity: days 3-7 as a cut foliage

Genus Name (Latin): *Schefflera*   Pronunciation: (shef-LEER-a)

Species Name (Latin): sp.

Common Name: **Variegated Schefflera**

Family Name (Latin and Common): *Araliaceae,* the Ginseng family

Related Family Members in Book: *Hedera*

Availability: indoor plant: year-round

Color: green with white variegation

Unique Characteristics: Leaf form is palmate; cut foliage is durable.

Design Applications: (medium) (form) Leaf form and interesting coloration adds accent. Use for wedding bouquets, or is suitable for smaller arrangements. Variegated leaf can brighten dark focal zones.

Longevity: days 3-7 as a cut foliage

Genus Name (Latin): *Senecio*   Pronunciation: (se-NEE-shee-o)

Species Name (Latin): *cineraria* 'Cirrus'

Common Name: **Dusty Miller**

Family Name (Latin and Common): *Asteraceae,* the Composite or Sunflower family

Related Family Members in Book: *Achillea, Ageratum, Artemisia, Aster, Bracteantha, Calendula, Callistephus, Centaurea, Chrysanthemum, Coreopsis, Cosmos, Craspedia, Dahlia, Echinacea, Echinops, Gerbera, Helianthus, Liatris, Rudbeckia, Solidago, x Solidaster, Tagetes,* and *Zinnia*

Availability: landscape plant: varies as an annual

Color: gray

Unique Characteristics: Leaves are gray and hirsute and work well as a pressed foliage; can also be useful as a cut foliage.

Design Applications: (small) (accent) Uniquely gray leaf adds interest to designs, especially in combination with bold colors. Plant works well in container gardens.

Longevity: varies

Genus Name (Latin): ***Senecio***

Pronunciation: (se-NEE-shee-o)

Species Name (Latin): *cineraria* **'Silver Dust'**

Common Name: **Dusty Miller**

Family Name (Latin and Common): *Asteraceae,* the Composite or Sunflower family

Related Family Members in Book: *Achillea, Ageratum, Artemisia, Aster, Bracteantha, Calendula, Callistephus, Centaurea, Chrysanthemum, Coreopsis, Cosmos, Craspedia, Dahlia, Echinacea, Echinops, Gerbera, Helianthus, Liatris, Rudbeckia, Solidago, x Solidaster, Tagetes,* and *Zinnia*

Availability: landscape plant: varies as an annual

Color: gray

Unique Characteristics: Color of gray is almost white; makes a great candidate for pressing; also can be used as a cut foliage.

Design Applications: (small) (texture) Gray leaf is nice for adding accent to container gardens, and works especially well in combination with

bold pink, red, and blue. Makes a good choice for pressed floral designs.

Longevity: varies

Genus Name (Latin): ***Senecio***

Pronunciation: (se-NEE-shee-o)

Species Name (Latin): *cineraria* **'Silver Lace'**

Common Name: **Dusty Miller**

Family Name (Latin and Common): *Asteraceae,* the Composite or Sunflower family

Related Family Members in Book: *Achillea, Ageratum, Artemisia, Aster, Bracteantha, Calendula, Callistephus, Centaurea, Chrysanthemum, Coreopsis, Cosmos, Craspedia, Dahlia, Echinacea, Echinops, Gerbera, Helianthus, Liatris, Rudbeckia, Solidago, x Solidaster, Tagetes,* and *Zinnia*

Availability: landscape plant: varies as an annual

Color: gray

Unique Characteristics: Leaves have a delicate, lacy appearance; makes a great choice for pressing; also can be used as a cut foliage.

Design Applications: (small) (texture) Lacey leaf with gray color is a nice accent for container gardens. Use as a dried foliage in pressed flower designs.

Longevity: varies

Genus Name (Latin): **Senecio** Pronunciation: (se-NEE-shee-o)

Species Name (Latin): *rowleyanus*

Common Name: **String-of-Pearls**, String-of-Beads

Family Name (Latin and Common): *Asteraceae,* the Composite or Sunflower family

Related Family Members in Book: *Achillea, Ageratum, Artemisia, Aster, Bracteantha, Calendula, Callistephus, Centaurea, Coreopsis, Cosmos, Craspedia, Dahlia, Echinacea, Echinops, Gerbera, Helianthus, Liatris, Rudbeckia, Solidago, x Solidaster, Tagetes,* and *Zinnia*

Availability: indoor plant: year-round

Color: green (light)

Unique Characteristics: Leaves are spherical and succulent on pendant stems; should be wired, picked, and taped before inserting into container arrangements.

Design Applications: (small) (texture) Round "bead" on thin stem is a personal favorite. Makes a wonderful choice for boutonnieres, corsages, or bouquets. Unique form adds tex-

tural interest. Use in contemporary designs to add a vertical line.

Longevity: varies as a cut foliage

Genus Name (Latin): **Setaria** Pronunciation: (se-TAY-ree-a)

Species Name (Latin): *italica*

Common Name: **Foxtail Millet**, Japanese Millet, Italian Millet

Family Name (Latin and Common): *Poaceae,* the Grass family

Related Family Members in Book: *Avena, Chasmanthium, Misanthus, Pennisetum, Phalaris, Phyllostachys, Sorghum,* and *Triticum*

Availability: commercial product: year-round

Color: tan (natural)

Unique Characteristics: Dried material has a fuzzy seed head on a wiry stem which provides physical movement; seed head available in other colors, especially nice is basil green.

Design Applications: (small) (texture) Dried material is most effective when bundled. Works well in combination with other grasses, wheat, barley, and millet. Provides an interesting texture when single stems are added to small-scale designs.

Longevity: long lasting

Genus Name (Latin): ***Solenostemon***

Pronunciation: (so-le-NOS-tee-mon)

Species Name (Latin): *scutellarioides*—**Wizard Series**

Common Name: **Coleus,** Flame Nettle, Painted Nettle

Family Name (Latin and Common): *Lamiaceae,* the Mint family

Related Family Members in Book: *Lavandula, Leonotis, Mentha, Molucella, Monada, Ocimum, Origanum, Perovskia, Physostegia, Rosmarinus, Salvia, Stachys,* and *Thymus*

Availability: landscape plant: varies as an annual

Color: multicolored

Unique Characteristics: Color combinations are widely varied with leaf margins and venation in striking combinations; many varieties available.

Design Applications: (medium) (accent) Leaf is a nice choice for drama. Use plant in shaded window boxes, urns, or landscape situations.

Longevity: varies

Genus Name (Latin): ***Sorghum***

Pronunciation: (SOR-gum)

Species Name (Latin): spp.

Common Name: **Broom Corn**

Family Name (Latin and Common): *Poaceae,* the Grass family

Related Family Members in Book: *Avena, Chasmanthium, Miscanthus, Pennisetum, Phalaris, Phyllostachys, Setaria,* and *Triticum*

Availability: commercial product: varies September-October

Color: black, brown, or tan shades

Unique Characteristics: Dried material has a unique seed head, and a pendulous habit; used for making brooms; also can be purchased as a perishable product, continuing to air-dry with some shrinkage of stem.

Design Applications: (small) (filler) Beaded appearance is a welcome addition of texture to autumnal displays. Use material for providing physical movement.

Longevity: long lasting

Genus Name (Latin): ***Stachys***   Pronunciation: (STAY-kis)

Species Name (Latin): ***byzantina***

Common Name: **Lamb's Ears,** Woolly Betony

Family Name (Latin and Common): *Lamiaceae,* the Mint family

Related Family Members in Book: *Lavandula, Leonotis, Melissa, Mentha, Molucella, Monarda, Ocimum, Origanum, Physostegia, Rosmarinus, Salvia, Solenostemon,* and *Thymus*

Availability: landscape plant: varies as a perennial

Color: gray

Unique Characteristics: Leaf is pubescent, having the appearance of a lamb's ear, and can be pressed.

Design Applications: (medium) (accent) Fuzzy-leafed plant provides accent and mass.

Longevity: varies

Genus Name (Latin): ***Stachys***   Pronunciation: (STAY-kis)

Species Name (Latin): **'Helene Von Stein'**

Common Name: **Lamb's Ears,** Woolly Betony

Family Name (Latin and Common): *Lamiaceae,* the Mint family

Related Family Members in Book: *Lamium, Lavandula, Leonotis, Melissa, Mentha, Molucella, Monarda, Ocimum, Origanum, Physostegia, Rosmarinus, Salvia, Solenostemon,* and *Thymus*

Availability: landscape plant: varies as a perennial

Color: gray

Unique Characteristics: Large, wonderfully pubescent leaf.

Design Applications: (large) (accent) Large fuzzy leaf provides visual mass and accent.

Longevity: varies

Genus Name (Latin): ***Strelitzia***   Pronunciation: (stre-LIT-see-a)

Species Name (Latin): *reginae*

Common Name: ***Strelitzia* Foliage**, Bird-of-Paradise Foliage

Family Name (Latin and Common): *Strelitziaceae*, the Strelitzia family

Related Family Members in Book: none

Availability: commercial product: year-round

Color: tan (natural)

Unique Characteristics: Dried material is also available in painted decorator colors; leaf stem, which adds considerable length, is not shown in photo.

Design Applications: (large) (line) Curly leaf adds unusual line and accent. Works especially well in high-style arrangements.

Longevity: long lasting

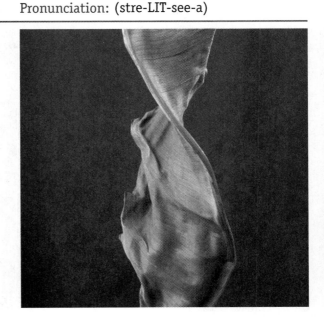

Genus Name (Latin): ***Swietenia***   Pronunciation: (swee-TEE-nee-a)

Species Name (Latin): *mahagoni*

Common Name: **Mahogany Pod**

Family Name (Latin and Common): *Meliaceae*, the Mahogany family

Related Family Members in Book: none

Availability: commercial product: year-round

Color: brown (natural)

Unique Characteristics: Dried material is also commercially available in painted colors on wood picks or wire stems.

Design Applications: (medium) (accent) Pod provides linear interest. Works well in combination with other cones and pods in mixed conifer or permanent botanical designs for holidays and everyday.

Longevity: long lasting

Genus Name (Latin): *Syngonium*

Pronunciation: (sin-GO-nee-um)

Species Name (Latin): *podophyllum*

Common Name: **Arrowhead vine,** Goosefoot

Family Name (Latin and Common): *Araceae,* the Arum family

Related Family Members in Book: *Anthurium, Caladium, Monstera,* and *Zantedeschia*

Availability: indoor plant: year-round

Color: green

Unique Characteristics: Leaf is often greenish white to cream; periodically appears as a commercial cut foliage.

Design Applications: (medium) (form) Arrow-shape leaf adds interest and mass. Makes a good choice for use in monochromatic white designs.

Longevity: days 3-5 as a cut foliage

Genus Name (Latin): *Thymus*

Pronunciation: (THY-mus)

Species Name (Latin): sp.

Common Name: **Thyme**

Family Name (Latin and Common): *Lamiaceae,* the Mint family

Related Family Members in Book: *Lavandula, Leonotis, Melissa, Mentha, Molucella, Monarda, Ocimum, Origanum, Physostegia, Rosmarinus, Salvia, Solenostemon,* and *Stachys*

Availability: landscape plant: varies seasonally

Color: green (medium)

Unique Characteristics: Plant is fragrant. Because of its diminutive size, this variety is often grown as an indoor plant, but can be used for outdoor gardens; other *Thymus* are available, some periodically appearing as a commercial cut foliage.

Design Applications: (small) (texture) This fragrant leaf works well with other herbs.

Longevity: days 3-5 as a cut foliage

Genus Name (Latin): ***Thuja***   Pronunciation: (THOO-ya)

Species Name (Latin): sp.

Common Name: **Arborvitae**

Family Name (Latin and Common):
Cupressaceae, the Cypress family

Related Family Members in Book: *Calocedrus, Chamaecyparis,* and *Juniperus*

Availability: commercial product: November-December

Color: green (medium to light)

Unique Characteristics: Flat sprays have scale-like surface; branch is also available preserved and dyed for use with permanent botanicals.

Design Applications: (large) (filler) Branches of flat sprays provide an excellent choice of texture.

Longevity: weeks 1-3 long lasting

Genus Name (Latin): ***Tillandsia***   Pronunciation: (ti-LAND-zee-a)

Species Name (Latin): ***usneoides***

Common Name: **Spanish Moss,** Graybeard

Family Name (Latin and Common):
Bromeliaceae, the Bromelia or Pineapple family

Related Family Members in Book: *Aechmea, Ananas,* and *Guzmania*

Availability: commercial product: year-round

Color: gray

Unique Characteristics: Dried or perishable product is available.

Design Applications: (small) (texture) Use to cover mechanics.

Longevity: long lasting

Genus Name (Latin): *Tillandsia*   Pronunciation: (ti-LAND-zee-a)

Species Name (Latin): sp.

Common Name: *Tillandsia,* Air Plant

Family Name (Latin and Common): *Bromeliaceae,* the Bromeliad or Pineapple family

Related Family Members in Book: *Aechmea, Ananas,* and *Guzmania*

Availability: indoor plant: year-round

Color: gray-green

Unique Characteristics: Leaves are narrow and curly; plant is epiphytic.

Design Applications: (small) (accent) Exotic appearance adds accent. Use in contemporary wedding bouquets or abstract arrangements.

Longevity: varies

Genus Name (Latin): *Tradescantia*   Pronunciation: (trad-es-KAN-shi-a)

Species Name (Latin): *fluminensis*

Common Name: **Wandering Jew**

Family Name (Latin and Common): *Commelinaceae,* the Spiderwort family

Related Family Members in Book: none

Availability: indoor plant: year-round

Color: green and purple variegated

Unique Characteristics: Coloration is unique; stem has vining habit.

Design Applications: (small to medium) (filler) Plant is useful for its trailing habit. Makes a good choice for mixed container gardens.

Longevity: varies

Genus Name (Latin): ***Triticum***   Pronunciation: (TRIT-i-kum)

Species Name (Latin): spp.

Common Name: **Bearded Wheat**

Family Name (Latin and Common): *Poaceae,* the Grass family

Related Family Members in Book: *Avena, Chasmanthium, Miscanthus, Pennisetum, Phalaris, Phyllostachys, Setaria,* and *Sorghum*

Availability: commercial product: year-round

Color: tan with dark beard

Unique Characteristics: This dried material is typically tan with beardless wheat species commercially available in several dyed colors.

Design Applications: (small) (texture) Seed head is great for bundling. Stems can also be used for binding technique. Makes a nice textural addition to fall centerpieces and wreaths.

Longevity: long lasting

Genus Name (Latin): ***Tsuga***   Pronunciation: (TSU-ga)

Species Name (Latin): ***canadensis***

Common Name: **Hemlock,** Canadian Hemlock, Eastern Hemlock

Family Name (Latin and Common): *Pinaceae,* the Pine family

Related Family Members in Book: *Abies, Larix, Picea, Pinus,* and *Pseudotsuga*

Availability: commercial product: limited November-December

Color: green (dark)

Unique Characteristics: Leaves are flattened and linear, soft to the touch; slender laterals along stem often have graceful habit.

Design Applications: (medium) (filler) Conifer has distinctive habit. Combines well with other mixed foliage for holiday arrangements or displays.

Longevity: weeks 1-3 long lasting

Genus Name (Latin): ***Tsuga***

Pronunciation: (TSU-ga)

Species Name (Latin): *canadensis*

Common Name: **Canadian (Eastern) Hemlock Cone**

Family Name (Latin and Common): *Pinaceae,* the Pine family

Related Family Members in Book: *Abies, Larix, Picea, Pinus,* and *Pseudotsuga*

Availability: commercial product: year-round with varying supplies

Color: brown

Unique Characteristics: This dried material is the smallest-scale cone listed in the book.

Design Applications: (small) (texture) Cone adds nice texture and interest to smaller-scale designs. Works well in combination with other sizes and shapes of cones, and is great for the paving technique. Use for holiday corsages, package tops, and ornaments.

Longevity: long lasting

Genus Name (Latin): ***Typha***

Pronunciation: (TY-fa)

Species Name (Latin): sp.

Common Name: **Cattail**

Family Name (Latin and Common): *Typhaceae,* the Cattail family

Related Family Members in Book: none

Availability: commercial product: year-round

Color: brown

Unique Characteristics: This dried material has several varieties available in varying sizes.

Design Applications: (small-large) (line) Spikes are a useful line element, especially in autumnal designs, and are a good candidate for bundling. Large sizes add linear mass.

Longevity: long lasting

Genus Name (Latin): ***Usnea***
Pronunciation: (YEW-snee-a)

Species Name (Latin): spp.

Common Name: **Florists' Lichen-Covered Branches**

Family Name (Latin and Common): *Parmeliaceae,* the Skull Lichen family

Related Family Members in Book: *Hypogymnia, Letharia*

Availability: commercial product: year-round

Color: gray

Unique Characteristics: Dried material; branches include scattered lichen, sometimes of several genera; photo depicts closeup of this genera.

Design Applications: (large) (line) Branches that are covered with lichens draw immediate attention. Large size works especially well as background or framing in naturalistic displays.

Longevity: long lasting

Genus Name (Latin): ***Vaccinium***
Pronunciation: (vak-SIN-ee-um)

Species Name (Latin): ***ovatum***

Common Name: **Huck**, Huckleberry

Family Name (Latin and Common): *Ericaceae,* the Heath family

Related Family Members in Book: *Erica, Gaultheria,* and *Rhododendron*

Availability: commercial product: year-round with varying supplies in July

Color: green (dark)

Unique Characteristics: Stem is woody with semiglossy ovate leaves; durability is an asset.

Design Applications: (small) (filler) Multi-leafed filler is useful in providing background or textural interest against *Chamaedorea* in traditional line-mass arrangements. Also is well suited for smaller-scaled designs.

Longevity: weeks 2-3 long lasting

Genus Name (Latin): ***Vinca***   Pronunciation: (VING-ka)

Species Name (Latin): *major* **'Variegata'**

Common Name: ***Vinca,*** Greater Periwinkle

Family Name (Latin and Common):
Apocynaceae, the Dogbane family

Related Family Members in Book: none

Availability: landscape plant: varies seasonally

Color: green margined with yellowish-white

Unique Characteristics: Vining plant is more
often sold as a bedding plant crop, but is
available in hanging basket form which is suitable for wedding and event decorations; typically not available as a cut foliage.

Design Applications: (small) (filler) Listed for
its extensive use in displays, the plant has
trailing vines which are a nice addition to
mixed container plantings.

Longevity: varies

Genus Name (Latin): ***Xerophyllum***   Pronunciation: (zee-ro-FIL-um)

Species Name (Latin): ***tenax***

Common Name: **Bear Grass,** Elk Grass, Indian
Basket Grass

Family Name (Latin and Common): *Liliaceae,*
the Lily family

Related Family Members in Book: *Asparagus,*
Aspidistra, Convallaria, Gloriosa, Hemerocallis,
Hosta, Hyacinthus, Kniphofia, Lilium, Liriope,
Muscari, Ornithogalum, Ruscus, Sandersonia,
Smilax, and *Tulipa*

Availability: commercial product: year-round

Color: green (medium)

Unique Characteristics: Plant air-dries well;
leaves are narrow and bladed, and display an
arching habit; *a personal favorite for adapt-*
ability and character.

Design Applications: (small) (filler) Multilinear
filler is an excellent choice for adding movement to container or bouquet work. Works well
in mass. Often is used to create loops in corsage and boutonniere designs.

Longevity: weeks 1-3 long lasting

Genus Name (Latin): *Zamia*

Pronunciation: (ZAY-mee-a)

Species Name (Latin): *floridana*

Common Name: **Coontie,** Florida Arrowroot

Family Name (Latin and Common): *Zamiaceae,* the Zamia family

Related Family Members in Book: none

Availability: commercial product: year-round

Color: green (dark)

Unique Characteristics: Leaves are spiny along rigid stem; provides a distinctive form.

Design Applications: (large) (mass) Linear leaf is an excellent choice for larger designs. Use with tropical flowers in contemporary design styles.

Longevity: weeks 1-2

Color Index

FLOWERS AND FRUITS

Color is crucial to design. It stirs emotion and provides drama. All commercially available flowers and fruits found in the book are listed below. They have been loosely sorted into color groups: red/pink, orange, yellow, green, blue/indigo, violet/purple, and white. These groupings include bicolors which are sorted by their dominant color. Use this list as a basis or refresher to aid in material selection within your color theme being created. Remember that product availability can vary by season; colors can vary by species and cultivar.

Red/Pink Hues

Achillea
Ailanthus
Alpinia
Alstroemeria
Amaranthus
Anemone
Anigozanthus
Anthurium
Antirrhinum
Arachnis
Astilbe
Bouvardia
Callistephus
Capsicum
Celosia
Centaurea
Chamelaucium
Chrysanthemum
Cleome
Cosmos
Costus
Cymbidium
Cytisus
Dianthus
Eremurus
Erica

Gerbera
Gladiolus
Gloriosa
Gomphrena
Heliconia
Hippeastrum
Hypericum
Iberis
Ilex
Ixia
Leptospermum
Leucadendron
Lilium
Limonium
Malus
Monarda
Nerine
Nigella
Paphiopedilum
Pentas
Phlox
Protea
Prunus
Rosa
Schinus
Sedum
Tulipa

Weigela
Xeranthemum
Zantedeschia
Zinnia

Orange Hues

Antirrhinum
Asclepias
Banksia
Bracteantha
Calendula
Carthamus
Celastrus
Chrysanthemum
Crocosmia
Dianthus
Euphorbia
Freesia
Gerbera
Gladiolus
Hippeastrum
Ixia
Leonotis
Lilium
Papaver
Physalis
Rosa

Sandersonia
Strelitzia
Tagetes
Zinnia

Yellow Hues

Acacia
Achillea
Alstroemeria
Anigozanthus
Antirrhinum
Bracteantha
Centaurea
Chrysanthemum
Craspedia
Cymbidium
Cytisus
Dianthus
Digitalis
Eremurus
Euphorbia
Forsythia
Freesia
Gerbera
Gladiolus
Heliconia
Iris

FOLIAGES AND DRIED MATERIALS

Another crucial design component is texture. Combinations of foliages or dried materials can produce strong textural effects within a design and play an important role. Varying forms, sizes, and colors can provide interest and mass. As a quick refresher for material ideas or to aid in color selection, use the following index. Commercially available items listed in the book are loosely grouped here by color categories: gray, green, variegated, and natural brown/tan. The perishable product color is listed here for foliages but keep in mind that many are available preserved and/or dyed into decorator colors. Dried materials are listed by photograph colors.

Gray

Acacia
Artemesia
Eryngium
Eucalyptus
Goniolimon
Leucadendron
Limonium
Perovskia
Rosmarinus
Stachys
Tillandsia

Green

Abies
Acer
Anthurium
Asparagus
Aspidistra
Atriplex
Buxus
Calathea
Callicladium
Callistemon
Camellia
Chamaecyparis
Chamaedorea
Cordyline
Cycas
Cytisus
Equisetum

Euphorbia
Fagus
Galax
Gaultheria
Hedera
Hosta
Humulus
Ilex
Laurus
Letharia
Liriope
Lycopodium
Magnolia
Mahonia
Melaleuca
Miscanthus
Monstera
Murraya
Myrtus
Nephrolepis
Papaver
Phormium
Pinus
Pittosporum
Platycerium
Podocarpus
Polystichum
Pseudotsuga
Ruscus
Sabal
Salix

Thuja
Vaccinium
Xerophyllum

Red/Yellow/Purple/Blue Hues or Variegated Colors

Acalypha
Acer
Atriplex
Aucuba
Bracteantha
Buxus
Caladium
Calathea
Codiaeum
Eucalyptus
Grevillea
Hedera
Hosta
Ilex
Lophomyrtus
Maranta
Miscanthus
Phormium
Phylaris
Phyllostachys
Physalis
Protea
Psylliostachys
Punica

Sabal
Triticum

Natural Brown/Tan

Abelmoschus
Avena
Betula (branch)
Eleusine
Euonymus (branch)
Lobaria
Lonicera (vine)
Nelumbo
Papaver
Phalaris
Picea (cone)
Pinus (cone)
Protea
Sabal
Salix (branch)
Setaria
Sorghum
Strelitzia
Swieteria
Triticum
Typha
Xerophyllum

Name Index

FOLIAGES AND DRIED MATERIALS